AF699765

Global Climate Change Linkages

Acid Rain, Air Quality, and Stratospheric Ozone

Global Climate Change Linkages

Acid Rain, Air Quality, and Stratospheric Ozone

Proceedings of a Conference Sponsored by
Center for Environmental Information, Inc., 99 Court Street
Rochester, New York 14604-1824

Edited by
James C. White, Ph.D.
Professor Emeritus, Cornell University

Associate Editors
William Wagner
Center for Environmental Information

Carole N. Beale
Center for Environmental Information

Elsevier

No responsibility is assumed by the publisher for any injury and/or damage to persons or property as a matter of products liability, negligence, or otherwise, or from any use or operation of any methods, products, instructions, or ideas contained in the material herein.

Elsevier Science Publishing Company, Inc.
655 Avenue of the Americas, New York, New York 10010

Sole distributors outside the United States and Canada:

Elsevier Applied Science Publishers, Ltd.
Crown House, Linton Road, Barking, Essex, IG11 8JU, England

© 1989 by Elsevier Science Publishing Company, Inc.

This book has been registered with the Copyright Clearance Center, Inc.
For further information, please contact the Copyright Clearance Center, Inc., Salem, Massachusetts.

This book is printed on acid-free paper.

Library of Congress Cataloging-in-Publication Data

Global climate change linkages : acid rain, air quality, and stratospheric ozone : proceedings of a conference / sponsored by Center for Environmental Information, Inc. ; edited by James C. White ; associate editors, William Wagner, Carole N. Beale.
p. cm.
"A compilation of papers presented at a conference in Washington, D.C. in late 1988, organized and sponsored by the Air Resources Information Clearinghouse (ARIC)"—Pref.
1. Climatic changes—Congresses. 2. Acid rain—Environmental aspects—Congresses. 3. Air quality management—Congresses. 4. Ozone layer depletion—Environmental aspects—Congresses. I. White, James C. II. Wagner, William, 1955- . III. Beale, Carole N. IV. Center for Environmental Information (U.S.) V. Air Resources Information Clearinghouse (U.S.)

QC981.8.C5G65 1989 89-39901
363.73'92—dc20 CIP
ISBN 0-444-01515-9

Current printing (last digit):
10 9 8 7 6 5 4 3 2 1

Manufactured in the United States of America

CONTENTS

Preface vii

About the Center for Environmental Information ix

Conference Organizing Committee ix

Conference Cosponsors xi

Introduction 1
Elizabeth Thorndike

Relationships Between Global Climate Change and Other Air Quality Issues 3
J. Christopher Bernabo

The Biggest Chill 13
Wallace S. Broecker

The Effects of Climate Change 23
Peter Robinson

Impact of Accelerated Sea Level Rise on Beaches and Coastal Wetlands 43
Stephen P. Leatherman

The Linkages Between Climate Change and Acid Rain 59
Hans C. Martin

Linkages Between Climate Change and Stratospheric Ozone Depletion 67
Robert C. Worrest, Katie D. Smythe and Alexander M. Tait

Linkages Between Global Climate Warming and Ambient Air Quality 79
Perry J. Samson, Stephen J. Augustine and M. Sanford Sillman

Climate Change, Smog and Acid Rain: Linkages Between Pollutants, Effects and Controls 95
Gordon J. MacDonald

Global Energy Use: A Quantitative Analysis 121
Richard Wilson

Global Climate Change Linkages: Strategies for Control of Emissions 165
Ian M. Torrens

Climate, Air Quality, Energy Alternatives and Efficiency 189
William R. Moomaw

The Role of Forests in Addressing the CO_2 Greenhouse 199
Gregg Marland

On the Benefits of Simultaneous Measurements — An Example 213
Lester Machta

Nitrogen Oxides and Ozone 219
Ralph M. Perhac

Policy Linkages 223
John Hoffman

Environmental Programs During an Era of Deficit Budgets 225
Lester B. Lave

The Environmental Attributes of Natural Gas 231
Michael I. German

Conference Participants 255

Conference Program 261

PREFACE

None of nature's phenomena are able to fluctuate without affecting, and being affected, by other phenomena. Current scientific opinion has it that we are in the grip of an unstoppable global temperature rise, brought on by human activity, which will reverberate in all directions, influencing many factors and involving a broad range of disciplines.

These diverse disciplines have not been in close communication, but now see themselves brought together by the linkages of their fields with global climate change.

This book is a compilation of papers presented at a conference in Washington, D.C. in late 1988, organized and sponsored by the Air Resources Information Clearinghouse (ARIC), which is a project of the Center for Environmental Information, Inc. (CEI).

The conference was organized to focus on relationships between potential climate changes and other atmospheric-driven issues, acid rain, air quality and stratospheric ozone.

For the reader, this book will provide an excellent assessment of probable results of climate change and of linkages with other atmospheric phenomena. Most important, the papers look seriously at options and strategies for dealing with climate change, outline research needs and examine possible national and international policies for coping with the future.

The broad interest in the meeting is evinced by the twenty-five diverse organizations and agencies which agreed to cosponsor the conference. We wish to extend our appreciation to the American Fishing Tackle Manufacturers Association, Electric Power Research Institute, Environment Canada, Motor Vehicle Manufacturers Association, Ontario Ministry of the Environment, Society of American Foresters and World Resources Institute for their support in helping to underwrite the conference. We wish to state that the opinions expressed are those of the authors and do not necessarily reflect the views of the Center for Environmental Information or the conference cosponsors.

We wish to thank the capable and distinguished program organizing committee, whose names are listed in these pages, for their exemplary role in identifying those people best able to clarify the issues. Special thanks go to J. Christopher Bernabo, whose ideas were the core of the program.

These proceedings are a cooperative effort of the staff of CEI. Special credit is due to Debra Segura for the onerous job of manuscript preparation and to the associate editors, William Wagner and Carole Beal, for the myriad details of technical editing and production.

This book brings the force of many disciplines to bear on a looming global problem. We hope it is a valuable resource for scientists and policy makers.

James C. White
Cornell University, June 1989

ABOUT THE CENTER FOR ENVIRONMENTAL INFORMATION

The Center for Environmental Information, Inc. (CEI) was established in Rochester, N.Y. in 1974 as an answer to the growing dilemma of where to find timely, accurate and comprehensive information on environmental issues. It is a private, non-profit organization funded by membership dues, fees, contracts, grants and contributions. The Center remains today a Rochester-based organization, but its services now reach far beyond the local community, reflecting the increasing number, scope and complexity of problems affecting the environment.

CEI's policy has always been to maintain a neutral position on environmental issues, in the belief that a nonpartisan information service removes communication barriers and helps individuals, public officials, businesses and organizations to be more effective in finding constructive solutions to difficult environmental problems. To meet this growing need for current and comprehensive information, CEI has developed a multi-faceted program of publications, educational programs and information services.

The Air Resources Information Clearinghouse (ARIC) is a project of CEI. ARIC gathers, organizes and disseminates information, and produces the *Acid Precipitation Digest* and the *Global Climate Change Digest*, which are published and distributed worldwide by Elsevier Science Publishing Company, Inc.

CONFERENCE ORGANIZING COMMITTEE

J. Christopher Bernabo
Science & Policy Associates, Inc.
Mohamed T. El-Ashry
World Resources Institute
John Miller
National Oceanic and Atmospheric Administration
Ralph M. Perhac
Electric Power Research Institute
Paul Ringold
National Acid Precipitation Assessment Program
James C. White
Cornell University Center for Environmental Research

CONFERENCE COSPONSORS

American Association for the Advancement of Science
American Fishing Tackle Manufacturers Association*
American Gas Association
American Geophysical Union
American Meteorological Society
Chemical Manufacturers Association
Climate Institute
Cornell University Center for Environmental Research
Edison Electric Institute
Electric Power Research Institute*
Environment Canada*
Environmental and Energy Study Institute
Environmental Defense Fund, Inc.
Motor Vehicle Manufacturers Association*
National Acid Precipitation Assessment Program
National Climate Program Office
National Council of the Paper Industry for Air and Stream Improvement
National Oceanic & Atmospheric Administration
National Wildlife Federation
Natural Resources Defense Council
Ontario Ministry of the Environment*
Quebec Ministry of the Environment
Society of American Foresters*
U.S. Environmental Protection Agency
World Resources Institute*

*Funding Support

Global Climate Change Linkages

Acid Rain, Air Quality, and Stratospheric Ozone

INTRODUCTION

Elizabeth Thorndike

Center for Environmental Information, Inc.
99 Court Street
Rochester, New York 14604-1824

I want to welcome you to this conference on Global Climate Change Linkages and, on behalf of the organizing committee, I would like to thank the 25 organizations, agencies and institutions in the U.S. and Canada which have endorsed this conference as co-sponsors and, in some instances, provided funding support. Both the co-sponsor and funding support come from a varied and representative spectrum of those concerned with this issue.

The problem of global climate change does not have a ready solution. There is no technological quick fix. However, just because a problem is global does not mean that all responses to mitigating the problem must be global. Climate change is the cumulative effect of myriad decisions about energy use and natural resource management made at the local and national levels. Dealing with the linkages between it and other air quality issues – ambient air pollution, acid rain and ozone depletion – will require changes from the grassroots up, as much as decrees from the top down.

The cumulative impact of our separate and collective use of fossil fuels must be assessed in terms of world energy needs, as well as in the context of an objective appraisal of the pros and the cons, the impacts and the methods of all forms of energy use and energy efficiency. This conference is one beginning.

RELATIONSHIPS BETWEEN GLOBAL CLIMATE CHANGE AND OTHER AIR QUALITY ISSUES

J. Christopher Bernabo

Science and Policy Associates, Inc.
Landmark Building – Suite 400
1333 H Street N.W.
Washington, D.C. 20005

Abstract

Climate is the most important environmental variable influencing ecological and societal systems. Climatic changes occur naturally on all time scales from decades to millennia. Man-induced climate forcing is superimposed on the background of natural fluctuations. The timing, magnitude, and patterns of future climatic changes are uncertain but human activities are releasing greenhouse gases that will play an increasing role in determining climate. Changes in climate affect the emissions of pollutants, atmospheric processes, and environmental impacts. These relationships link global climate change to other air quality issues such as acid deposition, attainment of ambient air pollution standards, and stratospheric ozone depletion.

Introduction

Climate is the primary environmental variable influencing natural systems and human activities. Changes in the extremes, variability, and mean conditions of climate will affect a series of existing air quality concerns including acid deposition, attainment of National Ambient Air Quality Standards (NAAQS) particularly for ground-level ozone, and depletion of stratospheric ozone.

These individual issues are linked by a complex set of relationships between emissions, atmospheric processes, and environmental consequences. Policies to address any one of these issues could influence the other problems because of the physical and societal linkages that exist among them. These linkages must be better understood in order to avoid undesirable side effects and to develop sound approaches to this ensemble of related issues.

Copyright 1989 by Elsevier Science Publishing Company, Inc.
Global Climate Change Linkages
James C. White, Editor

Naturally Changing Climate

Air quality regulation and other resource management decisions often are based implicitly on the *false* assumption that the climate (or average weather) is a constant. In fact, the Earth's atmosphere is highly variable on all time scales, from hours to decades and millennia or longer. The so-called "normal" weather of the television newscast is merely the average climatic conditions of the last 30 years and is not the mean state of a stable system. For various periods in the past, the 30-year means were very different. Inter-annual variations in weather are so large that they tend to act as noise obscuring the longer-term trends of climatic change.

Man's potential impact on climate needs to be understood relative to the varying natural climatic conditions. Figure 1 shows the history of climate over the last one million years derived from a variety of paleoclimatic indicators [1]. Throughout this period, the Earth has been in an "ice age" with a 100,000-year cycle of the waxing and waning of cold periods and the formation of huge ice sheets. Each of these glacial periods lasts about 90,000 years and is punctuated by shorter interglacial warmings of about 10,000 years each.

This glacial/interglacial cycle of our current Ice Age is controlled by changes in the Earth's orbital parameters that alter the distribution of incoming solar radiation [2]. These orbital changes are caused by the gravitational influences of other planets. The mechanisms by which the resulting slight variations in insolation cause dramatic oscillations of the Earth's climate are not understood. However, large changes in the concentrations of natural greenhouse gases are involved. The atmospheric levels of carbon dioxide (CO_2) and methane (CH_4) have risen and fallen in parallel with the global temperatures suggesting that major feedback processes exist, although we do not yet comprehend them [3].

We currently live at the end of a geologically brief warming in an otherwise cold epoch. Human civilization developed in the current interglacial period which began about 8000 BC. In the broad scheme of natural variations over the millennia, we are due to enter another glacial period [4]. The maximum warmth of this interglacial period occurred about 6,000 years ago (Figure 1c) when global temperatures were about 1°C warmer than the 1930 AD to 1960 AD mean.

On the shorter time scale, climate has constantly changed over the last 1,000 years (Figure 1b). A cool period known as the "Little Ice Age" (ca. 1550 AD to 1700 AD) had wide ranging impacts on human activities [5]. During the last 100 years, global mean temperatures have fluctuated significantly but an overall warming trend of about 0.5°C is evident (Figure 1a). This global warming probably is real even though some regions, such as the United States, may have shown no net change [6]. This is not contradictory because the United States represents less than 2% of the global surface area. With the great spatial variability in climate, all regions do not show uniform change and few, if any, will actually correspond to the global mean.

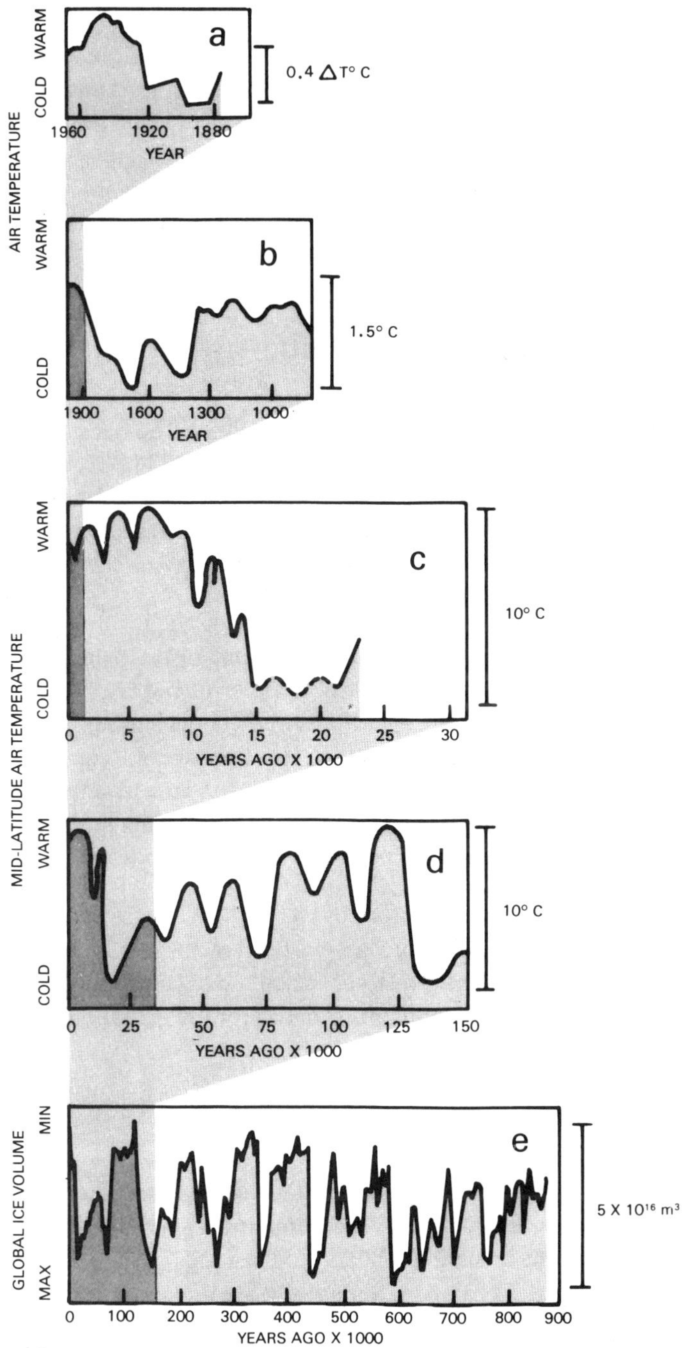

Figure 1. General trends in global climate for various time scales, ranging from decades to hundreds of millennia. Climatic change reconstructions are based primarily on records from a) instrumental data, b) historical information, c) pollen data and alpine glaciers, d) marine plankton and sea-level terraces, and e) isotopic fluctuations in deep-sea sediments [After: National Academy of Sciences (1975), Understanding Climatic Change, a Program for Action; 130p]

Nobody lives in the global mean. This number represents the relative state of the Earth's atmosphere and is not a predictor of specific regional conditions. The heterogeneous nature of the atmosphere is best understood by examining daily weather maps. Alaska, for example, can be experiencing a cold spell while the rest of the U.S. is very warm and Europe is moderate; but, when averaged globally, temperatures could be colder than "normal." The weather conditions in any given region are not necessarily indicative of the state of *global* climate. This fact will continue to confuse the public's understanding of what is happening to the Earth's climate.

Anthropogenic Greenhouse Forcing

Climatic change is an inevitable environmental reality. The question is: how, when and to what extent might human activities influence these naturally occurring variations? The climate change issue has been examined by scientists for over one hundred years. But the hot/dry summer of 1988 in parts of the United States catapulted it into the public's view and advanced the greenhouse issue to the top of the environmental policy agenda.

Greenhouse gases warm the atmosphere by allowing the incoming solar radiation to pass through like panes of glass, while trapping some of the outgoing infrared heat energy [7]. Greenhouse gases play a major role in determining Earth's climate and are responsible for keeping the surface temperatures warm enough to support life [8]. There is scientific consensus on several other matters: 1) the atmospheric concentrations of the major greenhouse gases are increasing; 2) emissions from human activities, such as the burning of fossil fuels, contribute to these increases; and 3) such increases would tend to raise the global mean temperature.

Unfortunately, this is not enough information on which to base predictions of future climates. To address the issue intelligently we must sort out what is known, what remains unknown, and where the major policy-relevant scientific uncertainties lie. Table 1 summarizes some of the major knowns and uncertainties for the greenhouse issue.

Emissions Relationships

The emissions of many air pollutants are climate sensitive. The amounts of SO_2, NO_2, and VOCs are influenced by climatic conditions in the source regions. For example, if a cold winter increases the demand for heating oil in a region, the related emissions of NO_x from combustion also would rise. Similarly, hotter summer temperatures cause greater demand for air conditioning and thus electricity; if coal is used to produce the electricity then SO_2 and NO_x emissions could increase.

A study in Minnesota found that a 1°C warming would raise peak summer electricity demand by about 10%, resulting in higher annual sulfur emissions in that region [10]. Similarly, the recent draft EPA report to Congress on the *Potential Effects of Global Climate Change in the United States* predicts that many regions will experience greater

TABLE 1

Summary of the Major Policy-Relevant
Scientific Knowns and Uncertainties
for the Greenhouse Issue

The degree of relative certainty associated with each known and unknown is indicated:

highly certain: ++
reasonably certain: +
relatively uncertain: -
highly uncertain: --

MAJOR KNOWNS

1. The atmospheric concentrations of greenhouse gases are a dominant factor determining global mean climate. *(++)*

2. The major greenhouse gases include CO_2, CH_4, O_3, H_2O, N_2O, and chlorofluorocarbons (CFCs). *(++)*

3. The concentrations of the major greenhouse gases (except possibly water vapor) have increased over the last 10 to 100 years. *(CO_2 - ++ ; Others - +)*

4. Human activities are significant contributors to the increases in the gases from the following anthropogenic sources:

   CO_2 Burning fossil fuels and land clearance. *(++)*

   CH_4 Rice cultivation, livestock, natural gas leaks, other agricultural and industrial processes. *(+)*

   O_3 Precursor gases: NO_x from combustion from stationary and mobile sources; VOCs fugitive emissions from liquid fuels, solvents and other industrial processes. *(+)*

   CFCs Synthetic compounds used for refrigeration, air conditioning, foam production, aerosol propellants, and solvents. *(++)*

   N_2O Fertilizers and fossil fuel combustion. *(+)*

5. Climate has changed due to natural causes throughout the last few thousand years and the influence of human activities are superimposed on these variations. *(++)*

MAJOR UNCERTAINTIES

1. The future concentrations of greenhouse gases probably will increase but the amounts are unknown. Projections must be based on uncertain assumptions about economics, technology, population, patterns of development, and other social factors. Current forecasts have an order-of-magnitude difference depending on the various assumptions used over the next 30 to 100 years. *(--)*

2. The magnitude of the climatic effect evident over the next 100 years is not known because of a) uncertainties in emission projections, and b) the lack of understanding of major feedbacks in the climate system such as the oceans, clouds, and natural emissions. About a four-fold variation in predicted future temperature changes is evident in models, depending on the assumptions about possible feedbacks. *(-)*

3. The timing of future climatic changes is uncertain by a factor of several decades or longer because of #1 above, plus lags in the ocean-atmosphere system, and the influence of natural events simultaneously affecting climate. Natural causes of climatic change are either unpredictable (e.g., volcanic eruptions) or unknown (e.g., the cause of the Little Ice Age). *(-)*

4. The regional patterns of any future climatic change are uncertain, especially for precipitation. Different global circulation models (GCMs) produce projections that do not agree even on the sign of precipitation change in broad regions [9]. Biological and societal effects result from the prevailing local conditions not the global mean situation. Although local and regional changes ultimately are related to the global mean, the specifics of these relationships are not yet well understood. *(Temp. - ; Precip. --)*

5. The effects of any future change in climate on natural and human systems is uncertain. Although a significant change in climate will have broad consequences, the specifics of the impacts are uncertain for two reasons: a) the patterns and rate of climatic change cannot be confidently predicted, and b) the climatic responses, particularly feedbacks, in natural and human systems are poorly understood. *(-)*

These knowns and uncertainties are not a comprehensive list. They are presented to highlight the status of some policy-relevant information about the greenhouse issue. The assigned levels of certainty and uncertainty are subjective judgments and are not statistically determined.

demand for electricity if the projected global warming occurs in the next 50 years [11]. This could increase emissions in those areas.

Transportation emissions (NO_x, VOCs, CO) also are climate dependent due to the effects of weather on combustion efficiency and air conditioning usage. VOCs released from fuel tanks, solvent use, and chemical plants are more volatile in higher temperatures; therefore, warmer conditions can cause elevated levels of VOCs. In fact, the majority of emission sources are affected by weather conditions, and thus are modulated by any changes in climate.

In addition to these traditional air pollutants (SO_2, NO_x, VOCs), the emissions of several major greenhouse gases are temperature dependent. For example, greater use of fossil fuels in response to warmer climate could result in more CO_2 emissions.

CFCs emissions not only can destroy stratospheric ozone but also they are potent greenhouse gases. Depletion of stratospheric ozone will increase the penetration of UV-B energy and could lead to further warming of the troposphere.

Natural sources of emissions also are sensitive to climatic conditions. The rate of CH_4 released from permafrost regions, nitrogen emitted by soil microbes, and VOCs derived from vegetation are all temperature dependent.

Atmospheric Processes Relationships

Climatic change, therefore, alters the basic weather features that control the transport, chemical transformation, and eventual deposition of all air pollutants. A region's climate is defined by the average condition of the meteorologic variables such as temperature, precipitation, humidity, cloudiness, and winds. When the Earth's climate changes, it results in shifting patterns, extremes, frequencies, and mean conditions of regional weather.

Stratospheric ozone depletion and tropospheric ozone pollution are linked through a series of complex atmospheric processes. A decrease in stratospheric ozone will result in an increase in UV-B radiation reaching the Earth's surface. This increase in UV-B irradiance will enhance such photochemical reactions as the production of tropospheric ozone and hydrogen peroxide, which also plays an important role in the formation of acid deposition. In addition, increased emissions of ozone precursors (NO_x, VOCs) from natural and man-made sources could further exacerbate the problem of ambient air quality.

The influence of prevailing weather on the patterns of air pollution can dominate over the distribution of sources. As shown in Figure 2, the annual pattern of ozone concentrations for the United States was very different between two consecutive years: 1978 and 1979. This condition did not reflect a shift in emissions, but rather was the result of different weather patterns.

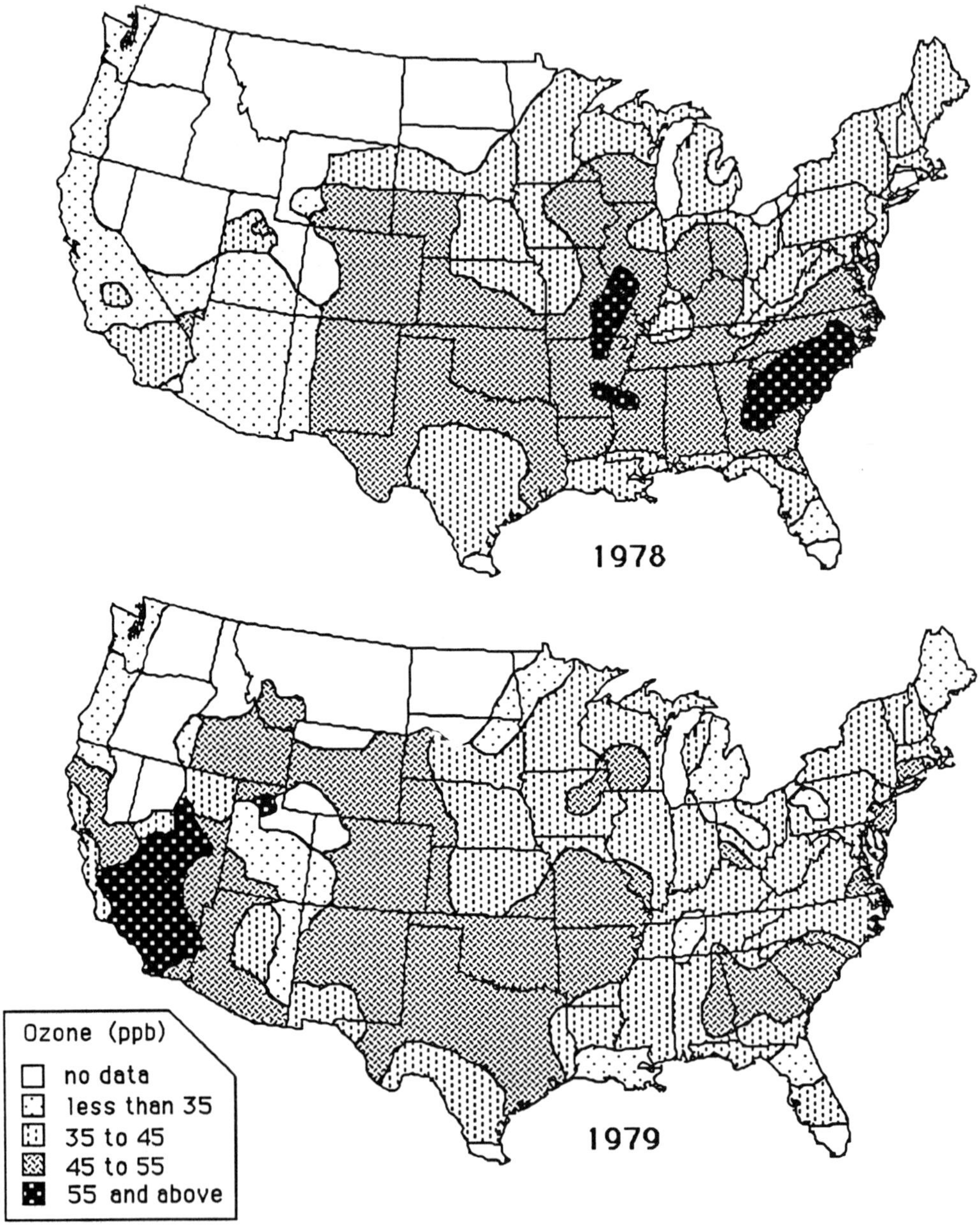

FIGURE 2. Effects of meteorology on pollution patterns: contrasting years of ground level ozone concentrations in the United States.

Overall, climatic changes will have significant effects on pollution levels because of the different types and frequencies of weather that will prevail in various regions. These complex atmospheric relationships are just beginning to be understood.

Environmental Effects Relationships

The effects of various air pollutants on human health, vegetation, aquatic systems, and man-made building materials are climate dependent to different degrees. Synergistic effects also may exist between the responses of different pollutants under various temperature and moisture conditions. For instance, the level of ozone that causes damage to a leaf is related to the amount of heat or moisture stress at the time of exposure. In addition, the sensitivity of certain materials to acid deposition and other air pollutants is influenced by the surface temperature and time of wetting. UV-B radiation also plays an important role in the degradation of certain materials.

To study the effects of pollutants on sensitive systems, researchers try to hold major climate-related variables constant. Understanding the complex interactions of climate- and pollution-induced stresses will require more sophisticated research and modeling. Simple extrapolation of pollution effects assuming a fixed climate may not suffice in the future.

Conclusions

Over the next century climate will change due to some combination of natural and anthropogenic causes. Man's activities are increasing the concentrations of greenhouse gases and that may produce a warming in mean global temperatures. However, the timing, magnitude, and environmental consequences of the anticipated global warming are still uncertain.

Global climate changes will influence man-made and natural emissions, atmospheric processes, and environmental as well as societal systems. Climate acts to modulate other air and water quality issues, each of which will be affected to varying degrees by changing conditions in the atmosphere. Acid deposition, attainment of national pollution standards for ground-level ozone, and depletion of stratospheric ozone each are linked to climate through temperature, precipitation, humidity, solar radiation, cloudiness, etc.

Past assumptions that climate is a long-term constant are incorrect and will lead to unreliable predictions of future pollution levels. Numerous air pollution - climate relationships exist and could result in complex and unexpected interactions in the future. A better understanding of the physical and societal linkages between these air quality issues is needed before the next generation of more holistic pollution control strategies can be developed.

The question of whether we have *enough* scientific certainty to undertake societal actions to address the issue is a value judgment not a technical conclusion [12,8]. There is no end point in science called regulate. The policy debate on what to do, how and when, therefore must include broad participation from governments, industries, environmental groups, international organizations, and the public.

References

1. J.C. Bernabo, Proxy Data: Nature's Records of Past Climates, *EDS Environmental Data Service Monthly*, N.O.A.A., U.S. Dept. of Commerce, March (1978).

2. J. Imbrie and J.Z. Imbrie, Modeling the Climatic Response to Orbital Variations, *Science* **207**, 943-953 (1980).

3. E.T. Sundquist, The Global Carbon Cycle and Quaternary Paleoclimates, *Episodes* **10**, 7-10 (1987).

4. G.J. Kukla and R.K. Matthews, When Will the Present Interglacial End? *Science* **178**, 190-191 (1972).

5. H.H. Lamb, *Climate Present, Past and Future, Vol. II* (Mathuem and Co., Ltd.) (1972).

6. K. Hanson, G.A. Maul, and T.R. Karl, Are Atmospheric "Greenhouse" Effects Apparent in the Climatic Record of the Contiguous U.S. (1895-1987)? *Geophys. Res. Letters* **16**, 49-52 (1989).

7. V. Ramanathan, The Greenhouse Theory of Climate Change: A Test by an Inadvertent Global Experiment, *Science* **240**, 293-299 (1988).

8. S.H. Schneider, The Greenhouse Effect: Science and Policy, *Science* **243**, 771-781 (1989).

9. S.L. Grotch, *Regional Intercomparisons of General Circulation Model Predictions and Historical Climate Data*, U.S. Dept. of Energy, DOE/NBB-0084 (1988).

10. E. Larson and D. Abrahamson, *Effects of Atmospheric Carbon Dioxide on U.S. Peak Electrical Generating Capacity*, Univ. of Minn. HHH/GEPP 83-2 (1983).

11. J.B. Smith and D. Tirpak, eds. The *Potential Effects of Global Climate Change on the United States, Draft Report to Congress* (U.S. Environmental Protection Agency, Washington, DC, 1988).

12. J.C. Bernabo, Science and Policy: Notes from a Former Congressional Fellow, *EOS*, Transactions of the American Geophysical Union **67**, 82 (1986).

THE BIGGEST CHILL*

When ocean currents shifted, Europe suddenly got cold. Could it happen again?

Wallace S. Broecker

Lamont-Doherty Geological Observatory of Columbia University
24055 Broecker
Palisades, NY 10964

We, the inhabitants of planet Earth, are performing a gigantic climate experiment. Begun by our grandparents, its results will be recorded by our grandchildren. The experiment involves the production and release into the atmosphere of gaseous molecules made up of three or more atoms; the most important of these are carbon dioxide (CO_2), methane (CH_4), and the freons (CF_3Cl and CF_2Cl_2). Unlike the two-atom molecules, oxygen (O_2) and nitrogen (N_2), which make up 99 percent of our atmosphere, these multi-atom molecules have the capacity to capture packets of outgoing radiation from the earth. Just as a blanket helps retain our body heat, these gases retain the earth's heat. Hence, the result of our experiment will be to make the surface of our planet warmer.

Unfortunately, our knowledge of the earth's climate system is still not good enough to reliably predict the effects of this heating on wildlife, agriculture, and a host of other matters important to humans. We will only know the results of the buildup of these "greenhouse" gases if our learning rate greatly accelerates.

In the face of such uncertainty, one might ask why the experiment is not declared dangerous to the well-being of the planet and abandoned. The reason is that the generation of greenhouse gases is not an enterprise designed by scientists. Rather, it is an inescapable by-product of our civilization. Carbon dioxide is produced when coal, oil and natural gas are burned. When carbon atoms, which make up the bulk of these fuels, combine with oxygen molecules from the atmosphere, an amount of CO_2 weighing roughly three times more than the fuel burned is generated. There is no feasible way to prevent this CO_2 from escaping into the atmosphere. Methane is produced by living organisms. The metabolic systems of steers and the bacteria in the mud of rice paddies are methane producers. Hence, some methane will be added to the atmosphere for each hamburger or bowl of rice we eat. Freons are manufactured by industry as foaming agents, refrigerants and propellants. Except for the freons, the greenhouse gases are products of activities essential to human survival. If five or so billion people are to be maintained on our planet, we must continue the greenhouse experiment. We are hooked.

*Reprinted with permission from *NATURAL HISTORY*, Vol. 96, No. 10; copyright the American Museum of Natural History, 1987.

Published by Elsevier Science Publishing Company, Inc.
Global Climate Change Linkages
James C. White, Editor

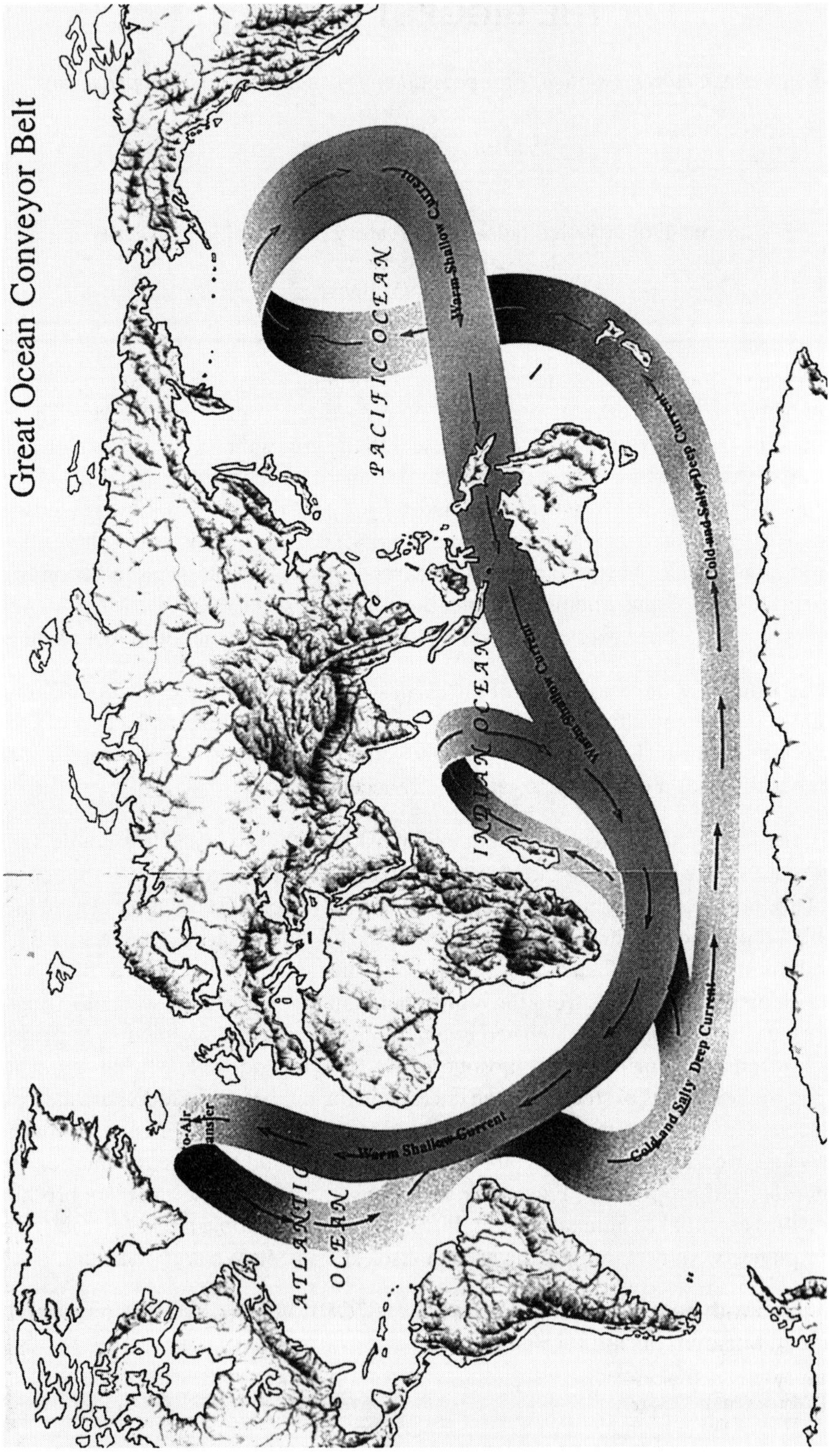
Great Ocean Conveyor Belt
PACIFIC OCEAN
INDIAN OCEAN
ATLANTIC OCEAN
Warm Shallow Current
Warm Shallow Current
Warm Shallow Current
Cold and Salty Deep Current
Cold and Salty Deep Current

Scientists struggle to increase our understanding of how the earth's environmental system operates in the hope that we will be able to predict at least some of the coming consequences. If so, we can develop strategies to cope with the "bad" and take advantage of the "good" results of this experiment.

These inquiries have recently revealed a piece of disquieting information. Geological studies suggest the earth's climate system resists change until pushed beyond some threshold; then it leaps into a new mode of operation. The situation is akin to that of a radio with automatic frequency control. When the dial on such a radio is turned, instead of one station fading out and the next one fading in, the radio remains locked on one station until a threshold is crossed, at which point it suddenly jumps to another station. The implication of this finding for future climates is clear: the effects of the greenhouse gas buildup may come in sudden jumps, rather than gradually. Such jumps would pose great threats to humans and wildlife.

Our suspicion that the earth changes in leaps comes from the evidence recorded in deep-sea sediments and in ice. The most studied of these records is the amount of heavy oxygen found in the preserved shells of microscopic animals on the ocean floor. The heavy form of oxygen in water vapor tends to be lost as atmospheric moisture is transported to the icecaps. The larger the icecap, the more heavy oxygen remains behind in the seawater. Thus, in eras when the icecaps were large, shelled organisms contained more heavy oxygen than they did when the icecaps were small; the shells therefore contain a history of the ice ages.

The oxygen isotope record tells us that over the last million years the polar icecaps have changed in a cyclic fashion, going from the rather small size of the current warm period to the very large size at the maximum of the last glaciation. More important, these fluctuations in ice volume have been shown the be in tune with periodic changes in the earth's orbit around the sun, generated by gravitational interactions among the objects making up our solar system. Because the timing of the oxygen isotope changes (as determined by age measurements on deep-sea sediment cores) matches what would be expected if the changes were driven by the earth's changing orbit, scientists are reasonably certain of the cause-and-effect relationship.

Why do changes in the characteristics of the earth's orbit have anything to do with climate? The answer is that these changes alter the earth's seasons. The relative amounts of each year's sunlight received during the winter months, as opposed to the summer months, changes in accordance with the changing orbit. Exactly how changes in the strength of the seasons drive the expansion and contraction of the earth's polar icecaps remains a matter of debate.

While the oxygen isotope record in the deep-sea sediments provided evidence pointing to the earth's orbital cycles as the pacemaker of glaciation, it also tended to lull scientists into concluding that the earth's climate responds gradually when pushed. This conclusion was drawn despite the realization that the response of polar icecaps to changing climate would have to be so sluggish that a smooth oxygen isotope record would be expected no matter how abrupt the changes in environmental conditions might be. So lulled were we that other clues in paleoclimatic records that pointed to abrupt response were largely disregarded.

The awakening came in the early 1980s when Hans Oeschger and his group at the university in Bern, Switzerland, carried out detailed measurements of the CO_2 content of air trapped in the ice from a deep boring made at a site in southern Greenland. These measurements concentrated on a section of the core on which earlier studies made by the Danish group of Willi Dansgaard had shown repeated leaps in Greenland's air temperature. To everyone's surprise, each of Dansgaard's jumps was accompanied by a 20 percent change in the CO_2 content of the air trapped in bubbles in the ice (and hence in the CO_2 content of air above Greenland at the time the ice formed).

Eyebrows were raised by Oeschger's CO_2 jumps because, while the temperature jumps could be written off as a curiosity of Greenland, the CO_2 changes could not. The atmosphere's CO_2 is well mixed with its other gases, hence a measurement in Greenland typifies the entire globe. Furthermore, the changes in CO_2 content found by the Oeschger group occurred in times as short as a few hundred years. To bring about these changes in CO_2 requires some extraordinary change in the earth's chemical cycles, particularly those operating in the ocean. Scientists were therefore forced to the realization that the leaps in Greenland's climate were far-reaching, involving the workings of the ocean as well as those of the atmosphere.

The new look at the ocean triggered by the finding of the Oeschger group brought to the fore the potential importance of a curious tie that exists between the functioning of today's ocean and today's atmosphere. This tie results in a globe-straddling ocean current that keeps northern Europe unusually warm. Paris lies almost a full ten degrees farther north than New York, yet its mean annual temperature is similar to that of New York.

The extra heat received by northern Europe is carried by a conveyor-belt-like ocean current. The part of the conveyor closer to the surface moves to the north; the conveyor's deeper part moves to the south. The important point is that the water of the upper part is warm, while that of the lower part is cold. The temperature change occurs at the northern limit of the belt (in the region around Iceland). Here, during the winter months, water warmed during its passage through the tropical and temperate Atlantic meets air cooled during its passage over frigid Canada. The meeting results in the transfer of heat from the sea to the air. The amount of heat is staggering, measuring about 30 percent of that received by the surface of the North Atlantic from the sun. The result of this transfer is twofold. First, the sting of the cold Canadian air masses is removed before the air hits northern Europe. Second, the waters are cooled and consequently made more dense. the extra density allows the water to sink to the abyss and feed the lower part of the conveyor. Thus the ocean current acts as a pump, extracting heat from low-latitude air and transferring it to high-latitude air.

The water that sinks to the bottom of the northern Atlantic flows down the full length of the Atlantic, around Africa, through the southern Indian Ocean, and finally up the

Pacific Ocean. This deep current carries twenty times more water than the combined world rivers.

There is also an ocean conveyor belt in the North Pacific but it runs the opposite way around. Deep waters move toward the north and upwell to the surface. From there they move toward the equator in the upper ocean. So in today's world, the Atlantic Ocean conveyor belt carries tropical heat for delivery to the atmosphere at high northern latitudes, while the Pacific conveyor belt carries cold surface waters southward, pushing the invading warm waters back toward the equator. Today's major ocean current system thus heats the lands adjacent to the northern Atlantic.

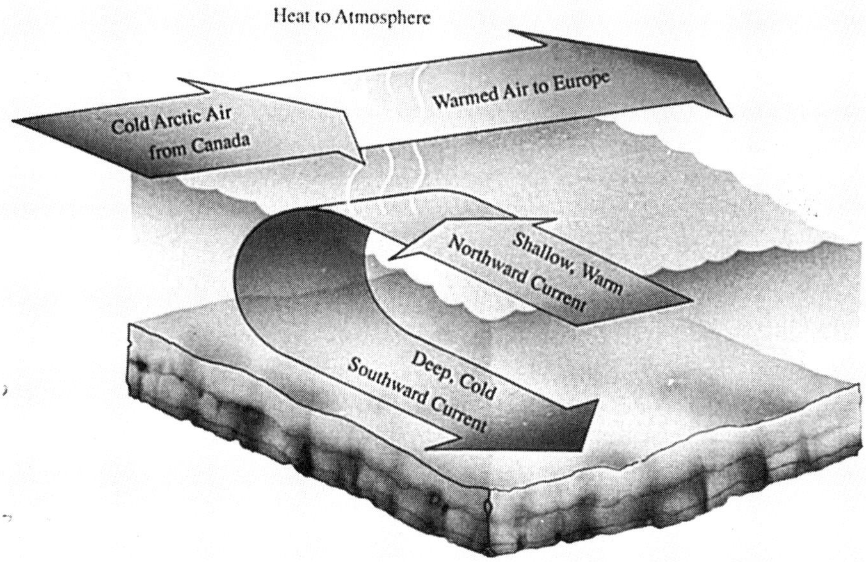

Sea-to-Air Heat Transfer *At the northern Atlantic limit of the ocean conveyor belt, surface waters release heat into the atmosphere, greatly moderating Europe's climate.*

While we don't have the complete answer to why our ocean operates in this fashion, we do have the first principles. The pattern of circulation is governed by the sea's salt. To understand this we must consider the transport of water through the atmosphere. The water that evaporates from the ocean falls eventually as rain or snow. Some of this precipitation reaches the land and some reaches the sea. Some precipitation that falls on land evaporates (mainly from plants) and some runs down the rivers and back to the sea. This cycle must exactly balance: for each molecule that evaporates from the sea, one molecule must return to it either by precipitation on its surface or from the mouth of a river. While this is true for the ocean as a whole, it need not be true for each part of the ocean. In fact, in today's world, an imbalance exists between the Atlantic and the Pacific. The Atlantic loses more water by means of evaporation than it gains by precipitation and continental runoff. The situation is reversed in the Pacific, which receives more water as rain and runoff than it loses by evaporation. While this imbalance is compensated for by a net flow of seawater from the Pacific to the Atlantic, it leaves a mark on the ocean's salt budget. Salt does not evaporate. Thus, the transport of water vapor from the Atlantic to the Pacific enriches the waters of the North Atlantic in salt content. The enrichment in salt must be compensated for by a flow of more salty water from the Atlantic to the Pacific. This is accomplished by the great conveyor belt: the water sinking to the abyss in the northern Atlantic carries excess salt.

The ocean conveyor system maintains higher surface water temperatures in the northern Atlantic than in the northern Pacific. Warmer waters have a higher vapor pressure and lose more water to the air by evaporation. Thus the rate of evaporation from the Atlantic is higher than that from the Pacific. This creates a global "still": water is extracted from the warm Atlantic and transferred through the atmosphere to the cool Pacific.

The phenomenon that maintains this situation is a devilish one; the circulation pattern is self-reinforcing and hence self-stabilizing. The deep current is driven by the extra density supplied to the waters of the northern Atlantic through the enrichment of salt. The enrichment of salt is driven by the heat carried by the water that flows northward in the upper Atlantic. Thus we have a classic chicken and egg situation; excess evaporation causes the deep current and the deep current causes excess evaporation.

The self-stabilization of this great conveyor belt is like the radio automatic frequency control already mentioned. And like that control, the mode of operation of the joint ocean-atmosphere operation will jump if pushed too far. The evidence contained in paleoclimatic records seems to be telling us that the conveyor of today's ocean did not function during the glacial time. Hence it is tempting to conjecture that the turning on and off of the conveyor constitutes an important link between the earth's orbits and our climate. When the belt is in operation, the warmth it delivers prevents ice from accumulating on the lands surrounding the northern Atlantic; when the conveyor is not in operation, these lands are sufficiently cold to permit their glaciation. If this is indeed the case, then the orbitally induced changes in seasonality must somehow alter the extent to which the water evaporating from the Atlantic Ocean escapes removal by the precipitation that falls on continental areas whose drainage is back into the Atlantic. Salt buildup is caused only by that fraction of the water evaporating from the Atlantic that escapes these basins and falls as rain in the Pacific or on continental drainage basins feeding the Pacific.

As we do not yet understand enough about the rules controlling the transport of water vapor through the atmosphere, we cannot say why changes in seasonality cause changes in the transport of water vapor from one ocean basin to another. We can only say that compelling evidence exists in the marine-sediment record for a flipping on and off of the ocean conveyor belt. Since the most vulnerable attribute of the conveyor is water vapor transport from the Atlantic to the Pacific, some link between this transport and seasonality seems logical.

Evidence for rapid jumps in climate on the land surrounding the northern Atlantic was discovered many decades ago by scientists studying pollen grains preserved in sediments. The record from bogs created during the early phases of the retreat of the icecap that covered Scandinavia and the British Isles during the last glaciation (20,000 to 14,000 years ago) shows a transition from the herbaceous shrubs of the cold period back to the forests of the warmer period. Those early postglacial forests persisted for about 2,000 years and then were suddenly replaced by shrubs akin to those of glacial time. This intense cold snap lasted about 700 years and then just as suddenly came to an end, permitting the forests to return. This brief reversion to cold conditions, which punctuated the period of deglaciation, was named the Younger Dryas (dryas is one of the herbaceous plants that clothed the landscape during the glacial time).

Like other signs pointing to rapid climate change, this rather extraordinary and relatively short-lived return to cold conditions was not given very high billing until Oeschger's group found the rapid CO_2 changes. It then became the focus of attention because detailed records for many localities on the earth's surface were available for the time interval when the earth emerged from its last episode of glaciation. Furthermore, the records from those sites showed that the Younger Dryas had a distinct geographic pattern. It is a prominent feature in records from the floor of the northern Atlantic and the surrounding continental areas (Maritime Canada, Greenland, and northern Europe). But it is not found in records from the continental United States.

This geographic pattern has been shown by computer simulation of the global climate system to be what would be expected if the ocean conveyor belt were to be turned off (and with it the enormous amount of heat delivered to the atmosphere above the northern Atlantic). Air temperatures over the eastern fringe of Canada, over Greenland, and over much of western Europe would plunge about 15°F. By contrast, no significant change in temperature would be seen in the United States. Hence, the geographic pattern of the Younger Dryas event points to the ocean conveyor as the causal villain.

Evidence from ocean-sediment cores taken in the northern Atlantic strengthens this case. Not only do the shells of the surface-dwelling plankton show a dramatic shift back to cold water species during the Younger Dryas interval but also, as shown by Edward Boyle of MIT, the content of the trace metal cadmium in the shells of the bottom-dwelling foraminifera indicates that cold deep waters, akin to those present during glacial times, returned during the Younger Dryas.

So we get the following picture. A full-fledged glaciation dominated our planet from about 20,000 to about 14,000 years ago. This corresponded to a time when the earth's orbital cycles gave rise to reduced seasonality at high latitudes in the Northern Hemisphere. During this time the ocean was "on another station," meaning its conveyor belt was not in operation. Then, as the orbits led to gradual strengthening of seasonality, the conveyor was switched back on. The heat released by the conveyor to the atmosphere over the northern Atlantic caused the great ice sheets to melt. But about 2,000 years later the conveyor came to an abrupt halt, bringing back cold conditions to the lands around the northern Atlantic. Then after a 400-year hiatus, the conveyor belt sprang back into action and has run steadily up to the present. Why the brief stoppage?

To understand what appears to be the answer to this question, we must first appreciate that a staggering 12-million cubic miles of water was bound up in the continental ice sheets of the Northern Hemisphere at the peak of the last glaciation. During the 5,000-year period when this ice melted away, the flow of meltwater must have averaged about half the current discharge of the Northern Hemisphere's combined rivers. Where it went would therefore have an important impact on the ocean's salt cycle.

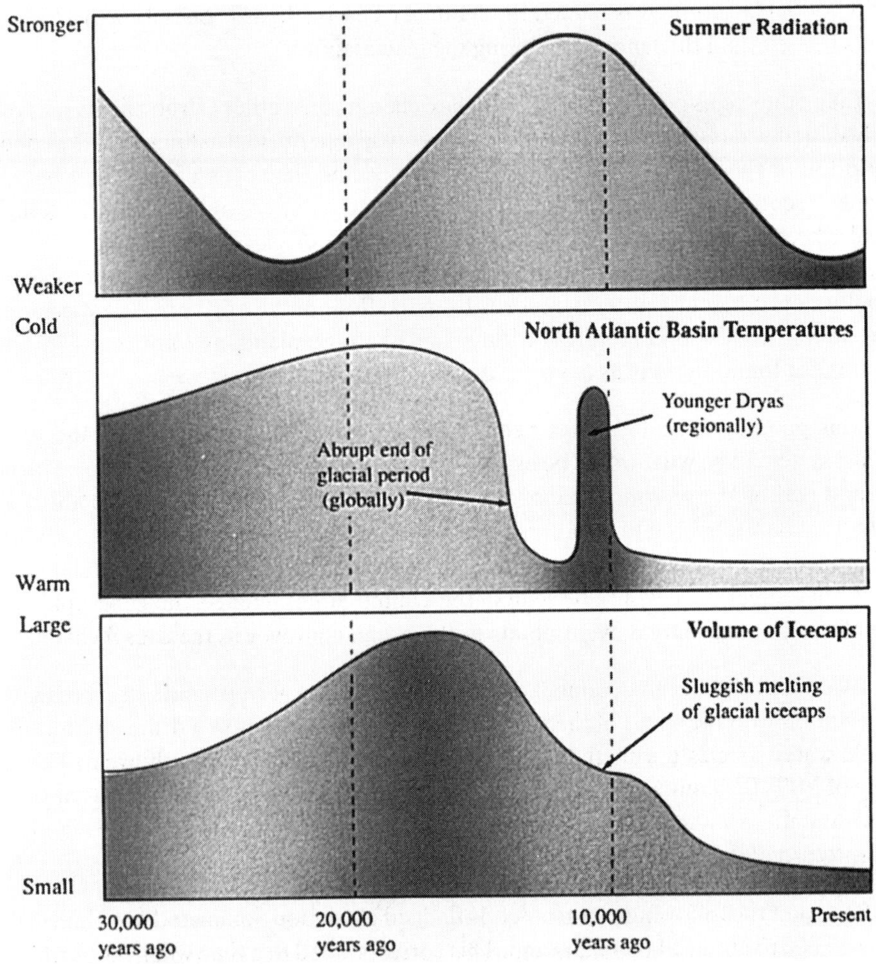

Fluctuations in the earth's orbit, affecting how much summer sunlight the earth receives (top graph), have expanded and contracted the polar icecap, driving the Northern Hemisphere in and out of glacial (Ice Age) episodes. Although the volume of the ice changes sluggishly and gradually (bottom graph), an abrupt global warming has marked the end of each glacial period, indicating that ocean–atmosphere operation can change suddenly (middle). The Younger Dryas event, a 700-year return to glacial conditions that interrupted the present warm episode, is thought to have been caused by a sudden diversion of melting Canadian ice sheet waters from the Mississippi to the Saint Lawrence River.

Geologists studying the deposits left behind during the retreat of the largest of the glacial icecaps, the one that covered Canada from the Rocky Mountains to the Atlantic Ocean, have reconstructed the routes taken by the meltwater released from its southern margin. Their studies reveal that, during the initial phases of melting, all routes converged on the Mississippi River and hence all meltwater flowed into the Gulf of Mexico. However, starting about 11,800 years ago, those routes were captured one after another by eastward-leading channels that progressively opened as the ice front retreated to the north. First, the water entering the basins of what are now Lake Erie and Ontario was diverted to the sea through the Hudson River valley. Then the water entering the basins of what are present-day Lakes Huron and Michigan was diverted to the Saint Lawrence River valley through a channel in the glacially depressed landscape north of Lake Erie. Finally, the biggest and most important of these diversions occurred when a lobe of ice blocking off the eastern end of what is now Lake Superior melted, allowing water to flow into Lake Huron and from there to the Saint Lawrence. At this point all the water melting from the southern margin of the Laurentide ice sheet was flowing eastward into the Atlantic. Nearly all of this water reached the Atlantic via the Saint Lawrence River. While the diversion of the meltwater from the Mississippi to the Saint Lawrence occurred in several discrete steps over a period of about 800 years, the largest of these diversions was the last, involving the meltwater released from about 60 percent of the southern perimeter of the Laurentide ice sheet. As dated by the radiocarbon method, it occurred about 11,000 years ago. Within the margin of error of such age determinations, this corresponds to the time the Younger Dryas began.

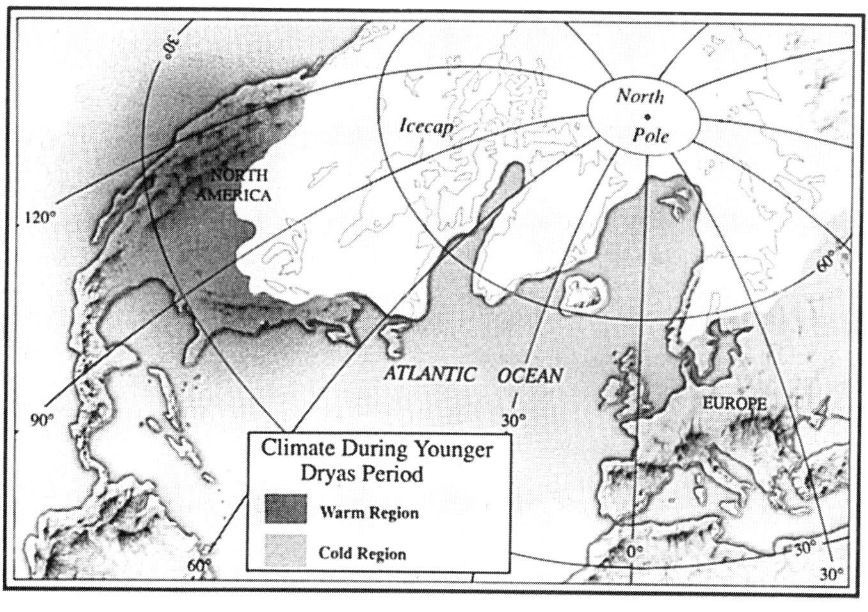

The sudden, localized cooling of Europe and Maritime Canada during the Younger Dryas event indicates that the great ocean conveyor belt suddenly stopped operating.

Scientists understand why the sudden diversion of a large meltwater flow from the Mississippi River to the Saint Lawrence would affect the ocean conveyor. As stated above, an essential feature of this system is the densification through cooling of high salt content water in the northern Atlantic. Today, the sinking of this water feeds a globe encircling deep current. The sudden influx directly into the northern Atlantic of an amount of water equivalent to that carried by the Amazon River would almost certainly diminish the salinity of surface waters in the northern Atlantic enough to disrupt the deep-water formation process. One might say that the Younger Dryas was not only a warning about the manner in which climate reacts when pushed, but also clearly showed that the coupling between the transport of fresh water across the earth's surface and the transport of salt within the sea is a critical element in the earth's response to climate change.

If this reading of the paleoclimatic record is correct, then we must face up to the reality that our climatic system does not operate in an orderly manner. As the greenhouse gases we produce build up, the ocean-atmosphere system may leap to yet another mode of operation. Unfortunately, the paleoclimatic record provides no clues as to how the earth's climate system responds when warmed beyond its prevailing state. Over the period for which our paleoenvironmental records are sufficiently detailed to permit such reconstructions, climate has not been significantly warmer than today's. So we are pushing the earth into an unknown realm. We have no way to predict how the great ocean conveyor will respond, nor can we be sure of other important elements of the system, which are subject to dramatic change. The computer simulations that have greatly improved our ability to predict weather have also told us some important things about the possible response to the greenhouse buildup. But because of their basic design, these computer models cannot tell us anything about the ocean-atmosphere system and leaping climates. At present no one knows how to incorporate the oceans into these simulations. Decades may pass before this can be done. Even then I doubt if computer simulations will offer much insight into the changes in climate that might be triggered by the greenhouse buildup.

The upshot is that we must take our greenhouse experiment more seriously. Rather than treating it as a cocktail hour curiosity, we must view it as a threat to human beings and wildlife that can be resolved only by serious study over many decades. We must expand our efforts to understand the operation of each of the units of the earth's climate system and how they interact. Only in this way will our grandchildren be able to prepare wisely for the changes that are bound to be wrought by our great experiment.

THE EFFECTS OF CLIMATE CHANGE

Peter Robinson

Department of Geography
University of North Carolina at Chapel Hill
CB 3220 Saunders Hall
Chapel Hill, NC 27599-3220

The effects of weather and climate on human activities are many and varied. They range from global, and highly significant, impacts in areas such as agricultural productivity to local and personal, and perhaps somewhat trivial, effects such as the delay in our airline flight departure because of fog. The effects may be directly and obviously associated with the atmosphere, as when a flood follows a severe storm, or rather indirect, as when a drought creates problems for hydroelectric power generation, forcing a utility company to import more expensive power, an action eventually reflected in an increase in our utility rates.

The effects are all primarily a reflection of the variability of our weather and climate. Until recently there was an unstated underlying assumption that this variability occurred around a mean condition which was more or less stable. Although we knew that there were long-term shifts in mean conditions, exemplified by our knowledge of the great Ice Ages, it was assumed that for practical purposes the mean conditions were essentially invariant. Concern could therefore be concentrated on the short-term variations, and we have developed coping mechanisms to deal with the anticipated day-to-day and year-to-year variability of weather and climate and its direct effects. Implicitly included in these coping mechanisms are strategies to deal with the indirect effects of climate. As a result, particularly in the developed world, we have proceeded under the assumption that particular climate events might create temporary problems in one sector of the economy, but that we could easily recover from them.

This assumption is immediately challenged when we consider climate changes associated with the greenhouse effect. There are clear indications that changes in the mean conditions and in the range of variability about that mean are likely. It is possible that the changes may be greater, and more rapid, than any seen before, creating extreme climates outside the range of any in human history. All parts of our planet will experience changes, although these will not necessarily be in the same direction all over the globe. Further, as a society we are becoming both increasingly sensitive to changes in weather and climate and increasingly part of a global community. Local climate problems can quickly become of global concern.

Copyright 1989 by Elsevier Science Publishing Company, Inc.
Global Climate Change Linkages
James C. White, Editor

Consequently the assumption that society is adjusted to climate variability is no longer valid. The familiar direct impacts of climate on various segments of our society will continue, but indirect impacts, stemming from the direct ones, are becoming increasingly important. Most direct impacts occur on a local scale, their effects spreading rapidly to regional, national and global indirect ones (Figure 1). While previous experience has given us some information about the direct impacts of the present climate, much less is known about the potential effects of a changed climate, while little is known about the indirect impacts. Thus specification of both the direct and the indirect effects of the changed climate on human affairs presents a major research challenge.

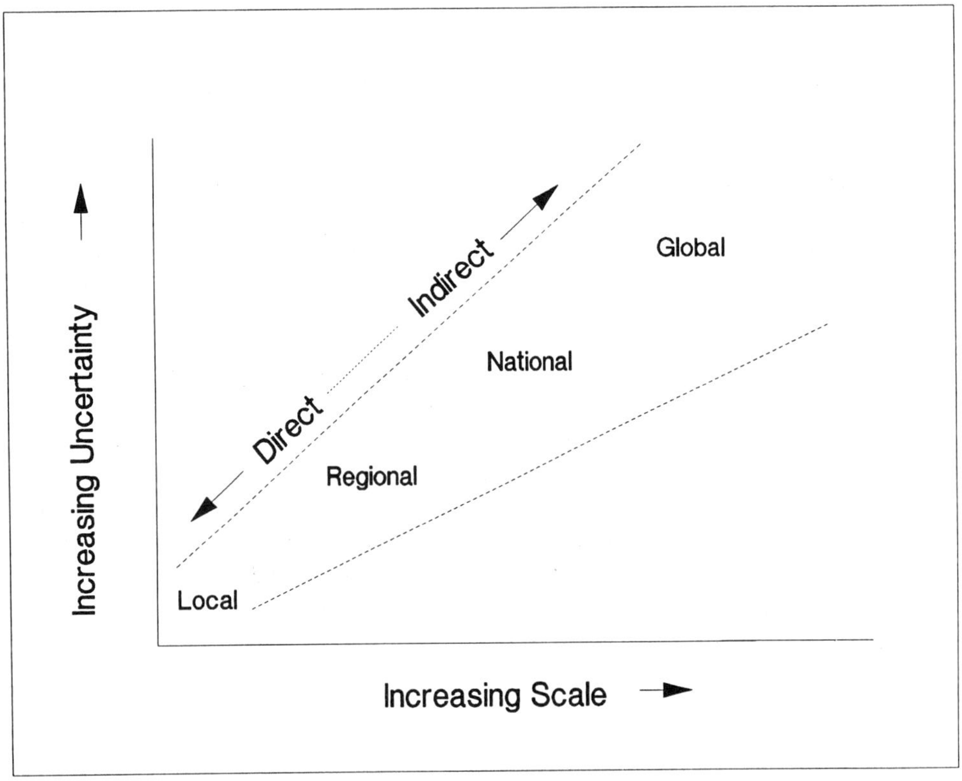

Fig. 1. A schematic view of the relationship between the temporal and spatial scales of impact of climate change and the current state of knowledge about the impacts.

Any impact assessment depends on an accurate depiction of the future climate. At the moment it is unclear what this will be like. However, it can reasonably be suggested that the changes will not be immediately apparent simply by walking about on the street. Instead we can expect changes in the frequencies of events. It may be, for example, that a warming in a particular area will be expressed as more hot days and fewer cold ones, although we may still complain about the cold from time to time. While new high temperature records may be set, we are unlikely to see all of our temperatures every day raised by (say) 10 degrees. If precipitation is to increase, we may have more rain days or an increase in precipitation from the storms we get, but again it is unlikely to be noticed that all of our storms suddenly become exceedingly violent. Similarly, we could have an increase in the frequency of droughts or hurricanes, the individual ones not being unprecedented, but in total creating a significant change and leading to a significant impact.

The future climate must be viewed in this rather local light, since any impact assessment must start with the local direct impacts. Unfortunately, our quantitative estimates of the future climate are best developed on the global scale, and are very tentative for local areas (Figure 2). The processes governing climate on most scales are qualitatively known, but cannot yet be incorporated into models to give reliable indications of future local climates. Indeed, unanticipated "surprises" are entirely possible. Hence the specification of the type of information needed for impact assessment presents a challenge to the climatologist.

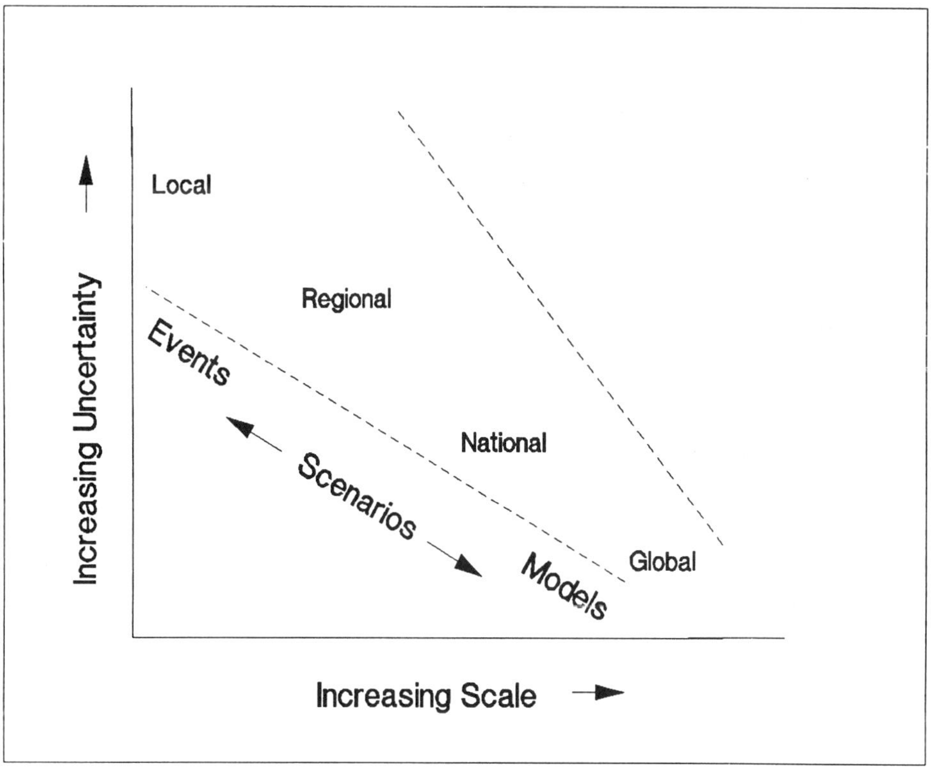

Fig. 2. A schematic view of the relationship between the temporal and spatial scales of climatic change and the current level of knowledge of possible future climates.

Figures 1 and 2 together illustrate the current dilemma in assessing the potential impacts of climatic change. Most of our knowledge of impacts is on a rather small scale, while our knowledge of future climates is best developed on the global scale. Research in all areas will undoubtedly refine our knowledge considerably. However, it is necessary to begin now to consider the development of appropriate policies to respond to the impacts of climatic change. Consequently there is a need to identify those areas where the impacts are likely to be the most significant and where policy development is most appropriate. Thus, in addition to those directly involved in assessing impacts or in climate research, experts in several other fields, ranging from economics and sociology to policy analysis, must be involved. Indeed, if we are to progress towards a rational policy, there must be strong links between these various disciplines (Figure 3). The following discussion, while placing greatest emphasis on the effects themselves, acknowledges throughout that links between disciplines must be developed and strengthened if viable policies for a response to climatic change, and the research necessary to produce them, are to be developed.

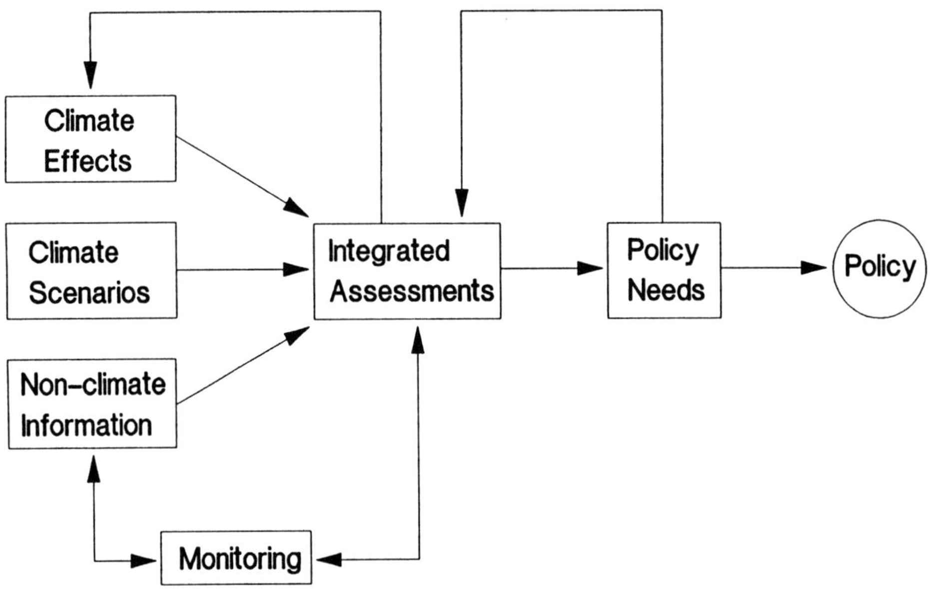

Fig. 3. The overall scheme of efforts required to assess the impacts of climate change and develop policy responses. After (Robinson and Hill, 1987)

Climate Effects

With so many possible effects on so many various aspects of human activity, it is impossible to deal with them all. Instead, possible impacts are considered for five areas: energy, agriculture, water resources, ecology and forestry, and air pollution and human health. No attempt is made to be comprehensive or exhaustive in the treatment of any of the areas. Rather, the focus is on typical problems. In most cases problems raised specifically in one area are equally relevant to each of the others. Deliberately missing from this analysis is consideration of the potential effects of sea level rise.

Energy

The range of direct impacts of weather and climate on various aspects of human activity can be illustrated by considering the utility industry (Table 1). Impacts cover a variety of time and space scales and many facets of the industry. Many of them are rather obvious, since they are impacts which occur under the present climate. The concern must be with changes in the frequency and severity of their occurrence. Taking ice storms as an example, for some utilities a changed climate may mean a disappearance of them, and the removal of a previous problem. For other areas, they may start to occur or increase in frequency, with implications to the utility for planning for a response to their occurrence.

TABLE 1
SOME DIRECT EFFECTS OF WEATHER AND CLIMATE ON UTILITIES

Time & Space Scale	Atmospheric Condition	Area of Effect
Short-term Local area	Thunderstorms	Power outages
	High winds	Transmission line problems Structural damage
	Ice Storms	Transmission line problems
	Heavy rain	Facilities damage Flood control problems
	Snowstorms	Maintenance crew activity Fuel transportation problems Stored hydroelectric water
	Tornadoes	General mayhem
	Hurricanes	Plant integrity
	Air Stagnation	Air quality compliance
	Heatwaves	Peak demands Plant operating efficiency Transportation problems
	Cold spells	High demand Fuel transportation
	Drought	Hydroelectric generation shortfall Little cooling water Barge fuel transportation problems
	Season temp.	Fuel usage
	Season clouds	Solar energy source
	Season winds	Wind power possibilities
	Season precip.	Water availability
	Annual mean	Rate changes
Long-term Large area	Long-term mean	Facilities planning

Such potential effects can be illustrated in detail by considering the demand for energy as a function of temperature. It has been estimated (McKay and Allsopp, 1980) that over 30% of North American and 50% of European energy is used to overcome the influence of climate. Most is for space heating and cooling, which accounts for much of the day-to-day variation in loads. Loads are sensitive to both high and low temperatures (Figure 4) so any change, and any change in variability, will have a direct impact on energy demand. The overall impact will vary from region to region, not only because of regional variations in the amount of climatic change, but also because of the relative shifts in the demand for heating or air conditioning. Using general circulation model results, Linder et al. (1987) projected an increase in annual and peak demand for utilities in both New York State and the southeastern United States, but the latter had a greater percentage increase because of a more significant increase in air conditioner use.

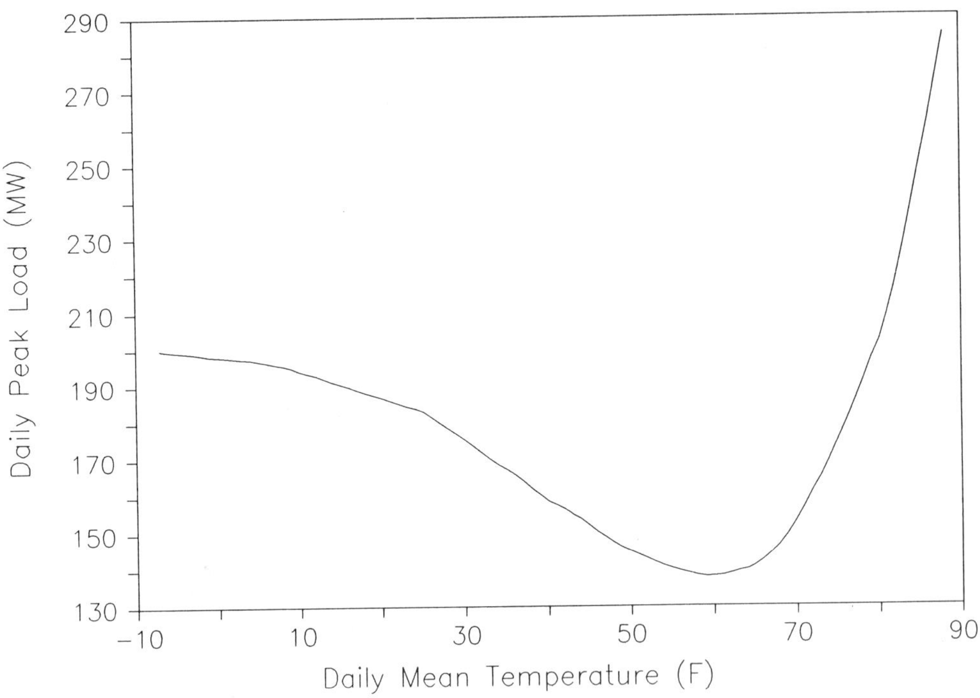

Fig. 4. The relationship between energy use and temperature for a typical electric utility in the mid-western U.S.A. Generalized from (Warren and LeDuc, 1981)

In addition to these direct heating and cooling needs, there are many other demands which in various ways are indirectly climate sensitive. It can be suggested, for example, that the energy required to pump irrigation water depends on the agricultural demand for water, which is largely climate driven. Similarly, climate influences the usage of various types of fuel oil. Climatic change may lead to changes in the desirability of various tourist venues, which may translate into a change in gasoline consumption patterns. Thus there is an ever widening sphere of influence as more and more indirect impacts are incorporated.

On the production side of energy, hydroelectric generation is directly climate sensitive. Linder et al. (1987), for example, foresee from their model projections significant decreases in hydroelectric production. Fossil fuel based generation, however, is indirectly climate sensitive, but there are a great number of facets to this. Even if we concentrate only on the possibility of the changed frequency of weather events, the range is enormous. The weather events may influence the ease of coal production or oil extraction by changing the frequency of adverse weather conditions. Strip mining and off-shore oil production are likely to be especially vulnerable. Transportation of fuel to the plant can be influenced by weather events. Increased cold spell frequency could produce more frozen railroad switches or longer periods when river ice prevents the movement of barge traffic. Alternatively, warmer conditions may lead to more frequent overheating and buckling of railroad tracks or low flows in rivers again preventing the movement of barge traffic.

Finally, the distribution aspect of the energy industry is not immune to climatic impacts. The possible direct role of events such as ice storms has already been mentioned. However, indirect considerations, such as regional shifts in demand as population shifts in response to climatic change, pose obvious problems. Here the emphasis must be on long-range planning by the utilities in order to have the capacity and the distribution network available to meet these changed needs.

Agriculture

The direct impact of climate on agricultural production is immediately obvious. Because it is so obvious and significant on a world-wide basis, it is appropriate to use this also as the vehicle to consider in some detail the potential indirect effects.

The direct impact of climate on agriculture is felt at the level of the field and the plant. Changes in factors, such as the length of the growing season, the frequency of hail, and the amount and timing of precipitation, all influence crop yields. Previous experience with year-to-year variability has led to the development of various crop-yield models relating climate and yield (e.g. World Climate Program, 1983) so that it is possible to estimate yields under a changed climate (Figure 5). Current estimates are based on the assumption that there is a shift in mean conditions, with the variability about the mean remaining essentially unchanged. This must represent a first approximation, since it is quite likely that the variability will increase dramatically, increasing the frequency of both very wet and very dry years. Thus the frequency of droughts may increase, an effect which demands a response if agriculture is to continue to be viable in an area. Possible responses (World Meteorological Organization, 1975) include changing crop types or cropping patterns, increasing irrigation or planting shelter belts. To cope with the other extreme, excess moisture, newly developed field drainage strategies may have to be implemented. The increased variability may also increase the number of heatwaves. This may create

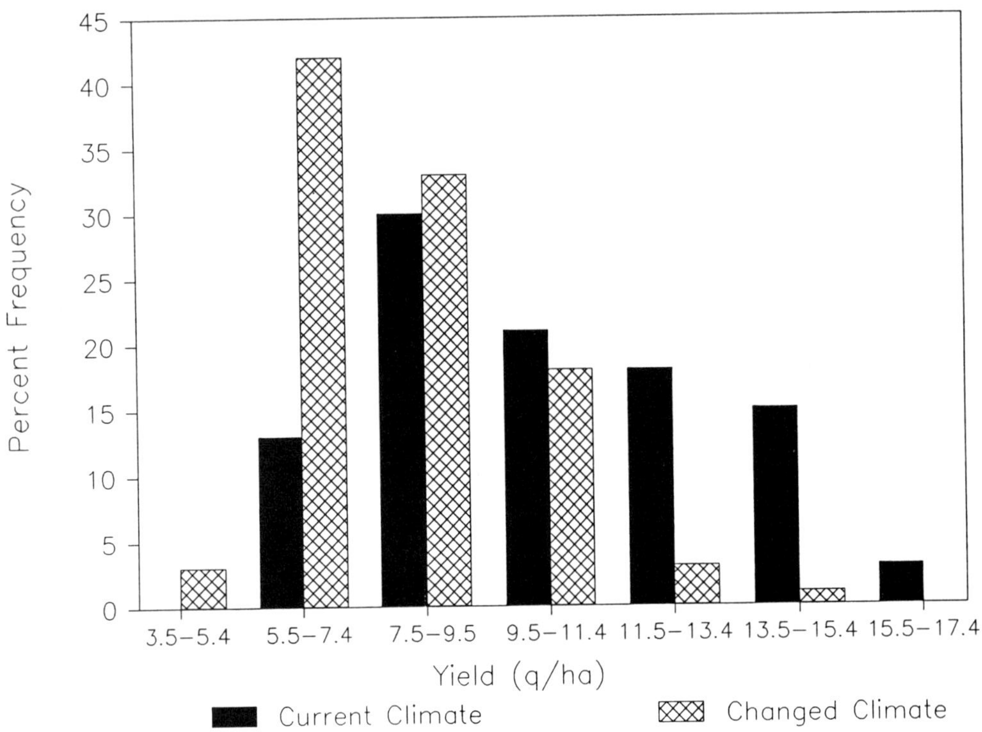

Fig. 5. Potential yields of spring wheat in North Dakota as a result of a 1°C increase in annual temperature and a 10% decrease in annual precipitation, assuming a similar annual variability to that of the present climate. Data from (Waggoner, 1983).

problems of heat stress for livestock, and may require the installation of air conditioned buildings if the animals are to survive.

An immediate implication of these effects is that additional resources are needed to maintain agricultural productivity. Impacts therefore begin to extend beyond the individual farm into the whole region, and eventually the whole world (Figure 6). Sales of irrigation equipment, farm equipment in general, and fertilizers are affected, influencing in turn regional cash flows and, on the broadest scale, industries such as automobiles, banking and chemicals. These industries may be located in areas far away from the region where the climate change is affecting agriculture. So we rapidly move from a very local impact to one with much wider implications, and an analysis of the impacts requires consideration of much more than just the agricultural effects themselves. However, a more thorough analysis is complicated by the fact that climate will be changing everywhere. Productivity may be increasing in some regions, decreasing in others. A full analysis of the impact, therefore, may start with simple climate-agriculture relationships, but rapidly requires the participation of other experts, whether economists to assess the links with all aspects of the economy, or sociologists to consider the social dislocations which might take place as the nature of the various agricultural regions changes.

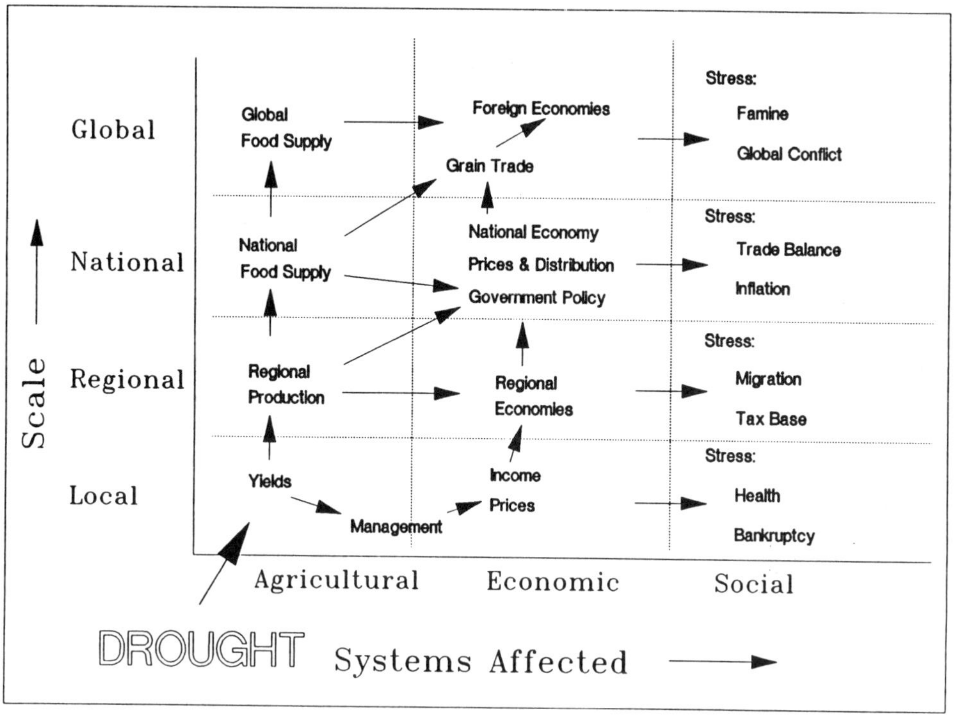

Fig. 6. Hypothetical pathways for the indirect impacts of climatic change on agriculture. Modified from (Warrick and Bowden, 1981)

The necessity for such a comprehensive analysis cannot be overemphasized. It is seductive to simply add up all the direct impacts to suggest that a particular country may be a "net winner" or a "net loser" under a particular postulated climate change. The potential for global conflict is immediately obvious. However, the fuller analysis incorporating the indirect impacts is likely to indicate few net winners. Indeed, the problems posed by climate change are likely to be global, requiring a coordinated and cooperative effort among all the nations of the planet (Atmospheric Environment Service, 1988).

Water Resources

Under the general rubric of water resources falls a whole range of issues. Indeed, some of the earliest work associated with the potential impacts of climate change were in this area (National Research Council, 1977). The issues involve both the amount and quality of water available for human use. Concern varies from considerations of flood magnitudes and frequencies, raising issues of flood control and floodplain planning, through low flow regimes, with associated water supply and water quality problems, to concern with long-term precipitation and its influence on ground water resources. The emphasis here is on how the effects and the climate information needs are a reflection of the different spatial scales of interest.

On the broadest scale the major climatic concern is with conditions averaged over a rather long period of time. This effect is relatively easily seen by considering variations in the level of the Great Lakes (Changnon, 1987). The Lakes primarily respond to variations in conditions integrated over long time periods, commonly years or decades, and for conditions integrated over the whole of the drainage basin. Individual storm events may have a major impact on a small sub-basin, but by the time the floodwaters enter one of the Lakes the effect is minor. The climate information needed to assess changes is similarly broad. Indeed, our current understanding of the regional effects of climatic change is beginning to be sufficient for tentative estimates of impacts on this scale. Thus Cohen and Allsopp (1988) used adjusted monthly average temperature and precipitation values determined from a general circulation climate model to determine long-term annual averages in their impact analysis. The model estimates the differences between the current climate and that when CO_2 is doubled, with results averaged over 4°lat. x 5°long. cells, a series of cells covering the Great Lakes drainage area. For water resources the results can be characterized by the change in net basin supplies, the inflow to the lake plus the precipitation onto the lake minus the evaporation from the lake (Table 2). Even with no change in demand, the less favorable conditions are obvious. When projections of increased demand, a result of population increase and economic development during the time that the climate is changing, are added, the picture becomes somewhat bleaker. The implications of shoreline falls provides an interesting counterpoint to the discussion of sealevel rise.

TABLE 2
PROJECTED EFFECTS OF CLIMATE CHANGE ON THE GREAT LAKES

Lake	Change in Net Basin Supply (%)		Change in level (m)
	Model Only	Model + Consumptive Use	
Superior	0	-0.7	-0.30
Michigan-Huron	-21.4	-29.2	-0.83
St. Clair	-17.9	-17.9	-0.73
Erie	-24.4	-63.7	-0.68
Ontario	-22.4	-33.8	*

* Cannot be calculated with present Lake operation policy

After Cohen & Allsopp (1988).

A somewhat similar approach has also been taken for the considerably smaller Sacramento River basin (Gleick, 1987). At this scale individual weather events can become important, seasonal differences are of major concern, and monthly average data are needed. On such scales climate model results must be treated with great caution, and it is most appropriate to suggest the impacts using a variety of possible future conditions (Figure 7). Because of the complex pathways of water as it moves from precipitation to runoff, some seasons are much more sensitive to climatic change than others. Certainly a particular percentage change in precipitation does not translate directly into a similar percentage change in runoff. Indeed, a somewhat similar but more general

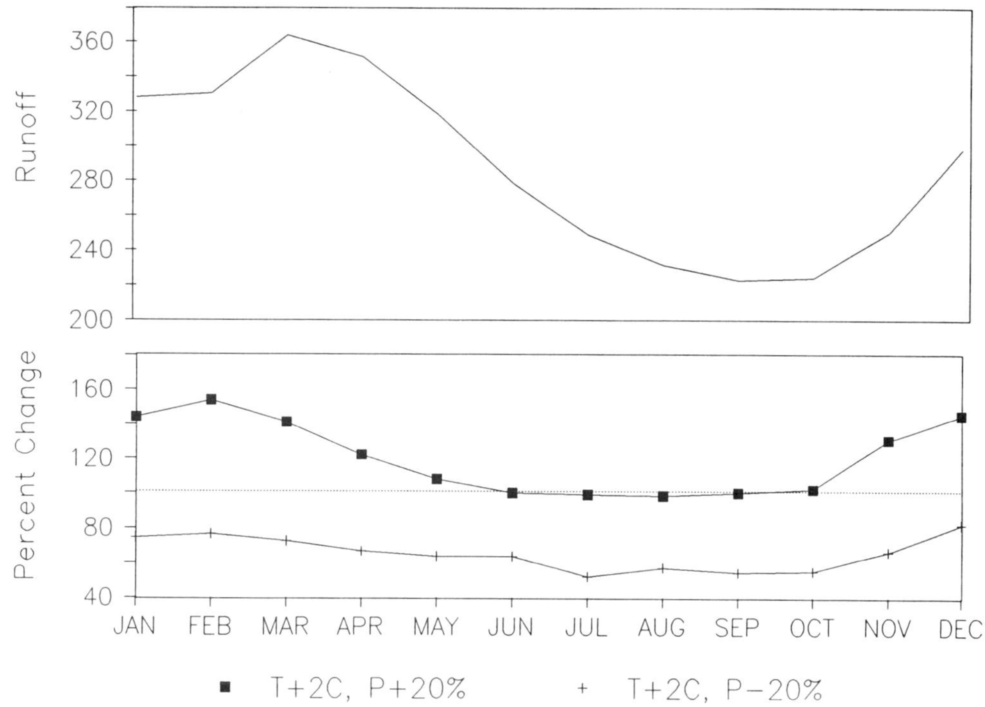

Fig. 7. Average monthly run-off for the Sacramento River in the present climate (above) and the percentage change in run-off for simulated climates with an annual temperature increase of 2°C and precipitation increases and decreases of 20%. Data from (Gleick, 1987)

analysis in humid climate conditions indicated the possibility that a 25% increase in annual rainfall could increase annual runoff by 50-70% (Nemec and Schaake, 1982).

For a small river basin, such as one in mountainous areas, the response is a rapid one to individual weather events. While long-term averages such as those appropriate for the Great Lakes may provide information on base flows within the streams in the basin, a much more important impact is likely to be the short-term variability of the flow. Most dramatic would be the frequency and magnitude of flood events, but a changed variability would have implications for dam design, planning for floodplain development and hydroelectric generation potential. Certainly even monthly average values are not appropriate here. Frequency distributions of weather events, or even indications of possible day-by-day weather streams, are needed if the likely impact, and possible responses, are to be addressed. However, this is the scale where our knowledge of climatic change is least developed.

Ecology and Forestry

That there is a close relationship between climate and natural vegetation has long been known. Indeed, much early climatological work was designed to provide a causal explanation of vegetation distribution. This culminated in the development of the concept of climate classifications and climate regions, exemplified by the classic Koppen scheme presented in all introductory climate texts. These classifications emphasized long-term mean conditions, implying that shifts in the mean lead to vegetation changes. More ecologically based work, however, demonstrates the important role of non-climatic factors in vegetation distribution. Soil type is one such factor. In the northern parts of the U.S. midwest for example, the soils still exhibit many characteristics they acquired during the Ice Ages which ended several thousand years ago, and the ecological character of this region still reflects this. Hence it is clear that ecological impacts will not be a simple "shift" in vegetation as the climate regions shift. Moreover, ecosystems are unlikely to "move" as an unchanged entity. The variable length and nature of life cycles of the many species involved dictates that they can move at various rates. Trees may advance at about 20 km per century, while insects, viruses and the like can move hundreds of kilometers a year (Roberts, 1988). Add to this the influence of inter-species competition and it is clear that changes in species composition will occur over and above any locational changes in ecosystems. This has profound implications for the biodiversity of the planet as climate changes. There is also concern that climatic changes may frustrate actions such as attempts to preserve wilderness areas, national parks, game preserves and the like.

Although such implications may be profound, the concerns are of recent origin, and relatively little is known about the possible ecological consequences of climatic change. On a fairly small, localized scale, there has been considerable success in understanding forest processes. Much of this understanding comes from the so-called gap models, where the focus is on competition and regeneration following the death of a tree which creates a clearing in the forest (Emanuel et al., 1985). An amalgamation of individual gaps to simulate the dynamics of the forest as a whole is beginning to provide the type of information needed if climatic impacts are to be adequately assessed. Such studies, of course, also have implications for the impact of pollution on ecosystems.

Another avenue which is beginning to be pursued is exploration of the role of fire in ecosystem dynamics. This is of concern in both forest and grassland ecosystems, where it is becoming clear that fire can play a major part in maintaining them. Most fires are of natural origin, resulting from lightning activity. Under a changed climate the frequency of thunderstorms, and thus lightning strikes, is likely to change. Moreover, the susceptibility of the vegetation to combustion is also likely to change, as the frequency of extended dry spells responds to the changed climate. Again, however, both climatological and ecological work is needed to identify the significance, and location, of such potential impacts.

Air Pollution and Health

The effect of the ozone hole and ozone depletion in general, with its implications for human health, has received great attention. While development of this hole is one result of human modification of the atmospheric composition, the broader results in the creation of climatic change may lead to impacts which are as, or more, dramatic. The Department of Energy "state-of-the art" report (White and Hertz-Picciotto, 1985) lists

14 areas where weather and climate can have a direct impact on human health. Most have simply been noted or postulated, with little research undertaken, so that very little is known for most of them.

One area has been explored in sufficient detail that the climate-health relationship is well established. This is heat stress, with a clear relation between high temperatures and mortality (e.g. Ellis et al., 1980). More than 200 Americans die annually from heat related causes (Morbidity and Mortality Weekly Report, 1981), and the number is much higher in years with major heatwaves. There is also an equal, but less explored, connection between heat stress and morbidity (e.g. Jones et al., 1982). While the connection is clear, it is not clear what constitutes the "threshold" above which stress starts to be felt. It is certainly a function of temperature and varies with region. One major factor is the number of sweat glands a person has, which is a function of the ambient temperature experienced within the first 2 years of life. In general those born in hot climates have more such glands and are better able to cope with heat (Kuno, 1956). Those moving to hotter climates can develop coping mechanisms but can never completely overcome their inherent lack of sufficient sweat glands. In the light of the suggestions of rapid temperature increases resulting from climatic change, this has implications for everyone. Comfort may become a problem for all, with some individuals having more severe problems. Even those people remaining in their native area may not be adapted to the new conditions. Since most problems arise when temperatures are well above normal, there must also be concern about the changes in the variability of temperature arising from climatic change. It is unclear at the moment how to determine the appropriate threshold condition before stress, illness or death occurs. There are some indications that a single very hot day may be sufficient, while other work suggests that an extended hot period is needed. It has also been postulated that overnight minimum temperatures are important, while the roles of humidity or the effects of moving into and out of air-conditioned environments have not been explored.

Of equal concern in the field of human health is air pollution. Here the effects depend on an aspect of climate which has traditionally been of little concern, the chemical climate. The meteorological situations which lead to high pollution concentrations will, under a changed climate, occur in the same way as at present, but their magnitude and frequency may change. In addition, however, the changed meteorological conditions may change the chemical climate through altered reaction or deposition rates (Table 3). These in turn may alter the physical climatic elements such as temperature, humidity and precipitation. At present this synergistic relationship is very poorly understood.

Scenarios of Future Climates

The above recital emphasizes the wide range of potential effects and the many unknowns in assessing them. A major objective of such assessments, of course, is the development of policies to anticipate or respond to them. This requires a concerted effort among a great variety of people (Figure 3).

TABLE 3
FACTORS OF CLIMATE CHANGE IMPORTANT FOR REGIONAL AIR QUALITY

Climate Element(s)	Influence
TEMPERATURE	Reaction rates
TEMPERATURE AND LIGHT LEVELS	Vegetation and soil emissions (hydrocarbons and NO_x) Deposition to vegetative surfaces
WATER VAPOR	Reaction rates (hydrocarbons, NO_x, O_3) Boundary layer dilution
AEROSOLS	Reaction rates Concentrations
PRECIPITATION	Deposition rates
CLOUD COVER	Reaction rates Conversion of SO_2 (acid deposition)
CIRCULATION PATTERNS	Local vs. remote deposition
FREQUENCY OF STAGNATION	Pollutant concentrations
MIXING LAYER HEIGHTS	Pollutant concentrations
STRATOSPHERIC OZONE	Reaction rates

After Penner et al. (1988).

A prime prerequisite for adequate assessments is the clear specification of future climates. Particular impacts may need very specific information about particular aspects of the climate. However, at present our knowledge of future climates is very general and imprecise (Figure 2). Certainly we cannot provide "predictions" in the same form as the familiar daily weather forecast. Instead the emphasis must be placed on the development of "climate scenarios" (National Research Council, 1982). These are sets of descriptions of future climates showing the range of conditions which could plausibly occur (Kellogg and Schware, 1981). Thus scenarios provide a variety of possible climates, all equally likely, the differences among them arising from the choices of input assumptions and forecasting methodologies made by the various producers. While they should incorporate the best scientific information and must be physically realistic, obeying the known laws of physics and chemistry which govern planetary processes, they depend greatly on the scientific insight and judgment of the scenario producer.

Scenarios developed for impact assessment require two major pieces of information. First is information about the climate processes likely to be acting in the future. These mainly come from the area of climate system research, notably from the "General

Circulation Models" of climate. These provide the indications of the broad-scale boundary conditions likely to be encountered in the future. They cannot at present provide the detail needed to undertake impact assessments. Hence, the second major piece of information must come through an analysis of past climates, whether the rather immediate past where we have detailed observational records or the more distant past where paleoclimatic techniques are revealing details of conditions during epochs when the climate was significantly different from the present day. Using past climate analysis, building on the boundaries specified by the models, we can begin to provide suggestions of future conditions with the detail needed to assess impacts.

The production of useful scenarios requires close cooperation between the climatologist actually producing the scenario and those responsible for assessing the impact. Care must be taken that the facets of the climate required for the assessment are adequately specified and that the scientific insight needed to produce information about these facets is used. Hence, there must be a strong and continuous link between the users and the producers.

This cooperation and the two-stage approach to scenario development have been used for almost all of the useful assessments produced so far. The degree of cooperation and the balance between the model and the analogue aspects of scenario development have varied. The works of Cohen and Allsopp (1988) and Gleick (1987) considered above are examples. In most cases the user of the scenario has been concerned with the direct impacts of climate. A further recent example, providing a comprehensive review of potential impacts, including some indirect ones, but with rather incompletely developed scenarios, is provided by the U.S. Environmental Protection Agency (Smith and Tirpak, 1988).

Non-climate Information

The close cooperation of the scenario producers and their immediate users will allow indications primarily of the direct impacts of climate. However, the indirect impacts may be as important, or even more so (Kates et al., 1985). As suggested above, assessment of the indirect impacts will involve experts in many diverse fields, including economists, sociologists, engineers, business leaders and politicians. They may, in turn, require the specification of other aspects of the climate, or indications of future climates in other regions. Hence, links between these people and the producers and the direct impact assessors are equally vital.

Two examples of the needs for links can be given. First, many disciplines mentioned above are likely to assess impacts using their own process models, whether they be econometric models simulating the national scene or hydrologic models relating precipitation to streamflow. Cooperation implies that the outputs of the climatologists' scenario "model" be readily usable as input to the user model. In particular, in addition to considering the same climatic elements, the spatial and temporal scales should match. Without this seemingly obvious, but difficult to achieve, meshing of models, there will be a lot of wasted effort and poor assessment of impacts. Second, equally obvious, but equally difficult, is the need for cooperation to ensure that the various components have similar accuracies. A tremendous amount of energy can be wasted producing a very detailed model for one aspect, only to find that the other aspects can be modeled very coarsely. It has been suggested, for example, that the climate models we have considered above may be more accurate than many of the econometric models to which they may be attached (Henderson-Sellers and McGuffie, 1987). Again close cooperation is the key.

Integrated Assessments

Although we have treated the individual effects more or less in isolation, in reality they cannot be so separated. Integrated assessments of the impact of climatic change on an area are vital if rational policies are to be developed. This is vital for two reasons. First, several impacts will occur simultaneously in any area. While one sector of a local economy may be suffering stress another may be benefitting. The net effect may be minor, requiring only minor changes in the social and economic structure of the area. Alternatively, of course, several major sectors may be severely stressed and major dislocation possible. Secondly, the impacts themselves may be physically synergistic. A climate change brings with it a suite of changes. A particular climate scenario will largely dictate the changes for each effect, and these must be treated as a suite if realistic assessments are to be made.

Monitoring

Although monitoring of climatic conditions is not a part of the assessment of impacts per se, it is a vital component of any system designed to assess and respond to climatic changes. The most obvious role is in the detection of climatic change itself. We need to know how the climate system is evolving. Since, as indicated above, the onset of climatic changes is likely to be subtle, without accurate detailed observations, no amount of analysis can detect the changes. And without this information it is unlikely that appropriate policy actions, however well founded, will be initiated at the appropriate time.

All scenarios of climatic change basically assume that the change will be a smooth, continuous one. Major concern is expressed that this smooth change may occur at an unprecedented rapid rate. For many reasons this is an assumption which we must make if we are to make progress in understanding and responding to the changes. There is, however, the possibility that the changes in the energy amounts and flows within the planet brought about by the greenhouse effect may create abrupt changes in climate. Thus we can postulate a "surprise-rich" future which is difficult to incorporate into scenarios. Early detection of any such abrupt change will be especially vital.

Although we have been thinking above of monitoring primarily in terms of the traditional climatic elements such as temperature and precipitation, there is another dimension from the impact assessment perspective. Cooperation between all those involved in the stages presented above may indicate that an entirely new type of observational program is required, one which may give a clearer indication of significant climatic changes, not from the purely climatic perspective, but from the perspective of significant impacts. Although it is impossible to specify now what this future monitoring need would be, such a concern is not unprecedented. Indeed, the whole history of concern with ozone depletion and acid deposition is permeated with the close linking of observational programs, atmospheric theory and potentially significant impacts.

Policy Needs

The ultimate end product of the various facets of impact assessment and its associated activities is the development of policies to mitigate adverse impacts and enhance beneficial ones. However, it is important to point out here that the policy needs developed in the policy arena must be communicated to those involved in developing integrated assessments and their component parts. Such feedback is vital. Without it, there is no guarantee that the research and monitoring needed for policy implementation will be undertaken in a timely and rational way (Chen and Parry, 1987).

Conclusions

Taken at its broadest, climatic change has the potential to affect every aspect of human lifestyle and all aspects of the surface and atmosphere of the planet. Many direct impacts of climate and its variability have been known for many years, and some relationships between climate and agriculture, water resources, energy, forestry and so on have been established. These impacts will continue, and the associated problems likely increase, under any future climate. Many investigations are underway and must continue, expanding our knowledge of these direct impacts. With a climatic change of the type foreseen to result from the greenhouse effect, however, a second set of impacts arise. These are the indirect impacts, which flow from the sectors feeling direct impacts into the whole of a regional economy and society and eventually into global impacts. Much less is known about these types of impacts although they are likely to be as important as, or more important than, the direct ones. Here there are only a few guidelines available allowing estimates of the magnitude and significance of these impacts, and basic research, developing assessment techniques as well as useful results, is vital.

For a comprehensive assessment of impacts, direct and indirect, which identifies the major sectors likely to be affected and which provides information relevant to policy formulation, experts from many disciplines are needed. Indeed, it is the forging of links between experts in many disciplines which provides the key to the successful development of indications of the major effects of climatic change and hence provides information guiding the future direction of the environment of the planet.

References

Atmospheric Environment Service, 1988, *The Changing Atmosphere: Implications for global security: Conference Statement*, Toronto, Canada, Government of Canada.

Changnon, S. A. Jr., 1987, "Climate fluctuations and record-high levels of Lake Michigan," *Bulletin of the American Meteorological Society*, 68, 1394-1402.

Chen, R. S. and M. L. Parry, 1987, *Policy-oriented impact assessment of climatic variations*, Laxenburg, Austria, International Institute for Applied Systems Analysis.

Cohen, S. J. and T. R. Allsopp, 1988, "The potential impacts of a scenario of CO_2-induced climatic change on Ontario, Canada," *Journal of Climate*, 1, 669-681.

Ellis, F. P., H. P. Prince, G. Lovatt and R. M. Wittington, 1980, "Mortality and morbidity in Birmingham during the 1976 heat wave," *Quarterly Journal of Medicine*, 49, 1-8.

Emanuel, W. R., H. H. Shugart and M. P. Stevenson, 1985, "Climatic change and the broad-scale distribution of terrestrial ecosystem complexes," *Climatic Change*, 7, 29-43.

Gleick, P. H., 1987, "Regional hydrologic consequences of increases in atmospheric CO_2 and other trace gases," *Climatic Change*, 10, 137-161.

Henderson-Sellers, A. and K. McGuffie, 1987, *A climate modelling primer*, New York, John Wiley and Sons.

Jones, T. S., A. P. Liang, E. M. Kilbourne, M. R. Griffen, P. A. Patriarca, S. G. Wassilak, R. J. Mullan, R. F. Herrick, H. D. Donnell, K. W. Choi and S. B. Thacker, 1982, "Morbidity and mortality associated with the July 1980 heat wave in St. Louis and Kansas City, MO.," *Journal of the American Medical Association*, 247, 3327-3331.

Kates, R. W., J. H. Ausubel and M. Berberian (eds), 1985, *Climate Impact Assessment*, New York, John Wiley and Sons.

Kellogg, W. W. and R. Schware, 1981, *Climate Change and Society: Consequences of increasing atmospheric carbon dioxide*, Boulder, CO, Westview Press.

Kuno, Y., 1956, *Human perspiration*, Springfield, Charles C. Thomas.

Linder, K. P., M. J. Gibbs and M. R. Inglis, 1987, *Potential impacts of climate change on electric utilities*, Report 824-CON-AEP-86, ICF Incorporated, for: New York State Energy Research and Development Authority, Edison Electric Institute, Electric Power Research Institute and U.S. Environmental Protection Agency, Washington DC, ICF Incorporated.

McKay, G. A. and T. Allsopp, 1980, "The role of climate in affecting energy demand/supply," in: Bach, W., J. Pankrath and J. Williams (eds), *Interactions of energy and climate*, Reidel, Dordrecht, The Netherlands, 53-72.

Morbidity and Mortality Weekly Report, 1981, editorial: "Heat wave related mortality – United States," *Morbidity and Mortality Weekly Report*, 29, 357-359.

National Research Council, 1977, *Climate, climatic change, and water supply*, Washington DC, National Academy Press. 132 pp.

National Research Council, 1982, *Carbon dioxide and climate: A second assessment*, Washington DC, National Academy Press.

Nemec, J. and J. Schaake, 1982, "Sensitivity of water resource systems to climate variation," *Journal of Hydrologic Science*, 27, 327-343.

Penner, J. E., P. S. Connell, D. J. Wuebbles and C. C. Covey, 1988, *Climate change and its interactions with air chemistry: Perspectives and research needs*, Livermore, California, Lawrence Livermore National Laboratory.

Roberts, L., 1988, "Is there life after climatic change?" *Science*, 242, 1010-1012.

Robinson, P. J. and H. L. Hill, 1987, "Toward a policy for climate impacts," *Bulletin of the American Meteorological Society*, 68, 769-772.

Smith, J. B. and D. Tirpak, 1988, *The potential effects of global climate change on the United States: Report to Congress*, U.S. Environmental Protection Agency, Office of Policy, Planning and Evaluation, Office of Research and Development (in preparation).

Waggoner, P. E., 1983, "Agriculture and a climate changed by more carbon dioxide," in: National Research Council: *Changing climate: Report of the carbon dioxide assessment committee*, Washington DC, National Academy Press. 383-418.

Warren, H. E. and S. K. LeDuc, 1981, "Impact of climate on energy sector in economic analysis," *Journal of Applied Meteorology*, 20, 1431-1439.

Warrick, R. and M. Bowden, 1981, "Changing impacts of drought in the Great Plains," in: Lawson, M. and M. Baker (eds.), *The Great Plains, perspectives and prospects*, Lincoln, Nebraska, Center for Great Plains Studies.

White, M. R. and I. Hertz-Picciotto, 1985, "Human health: Analysis of climate related to health," in: White, M. R. (ed), *Characterization of information requirements for studies of CO_2 effects: Water resources, forests and human health*, Washington, D.C., United States Department of Energy, Office of Energy Research, DOE/ER-0236.

World Climate Program, 1983, *Guidelines on crop-weather models*, Geneva, World Meteorological Organization, WCP-50.

World Meteorological Organization, 1975, *Drought and Agriculture*, Geneva, World Meteorological Organization, WMO-392, Technical Note 138.

IMPACT OF ACCELERATED SEA LEVEL RISE ON BEACHES AND COASTAL WETLANDS

Stephen P. Leatherman

Laboratory for Coastal Research
1113 LeFrak Hall
University of Maryland
College Park, Maryland 20742

Abstract

A significant portion of the world's population lives within the coastal zone, with many buildings and facilities built at elevations less than ten feet above sea level along the shoreline. Presently, many of these structures are not adequately above existing water levels or located far enough landward to ensure their survival and the safety of residents during major storm activity. This hazard has grown increasingly apparent and serious along much of the U.S. east and Gulf coasts, particularly the highly urbanized sandy barriers, as relative sea levels have risen during the twentieth century.

The general effect of sea level rise on coastal lowlands is to induce landward retreat – beaches erode and marshes are lost. Most sandy shorelines worldwide have experienced recession during the past century. Such has also been the case along the U.S. coasts; historical data indicate that about 90 percent of our sandy beaches are eroding. Accelerated sea level rise due to greenhouse-induced global warming will only increase erosion rates and exacerbate the present shoreline dilemma. Also, coastal marshlands will be lost at ever-increasing rates, with the nation's marshes perhaps disappearing in the coming decades at the high rates of loss presently experienced in Louisiana.

Introduction

Changing sea levels have been the driver of shorelines over geologic time. Tens of millions of years ago, the seas had advanced over a hundred miles farther inland than today; these old shorelines are marked by the "falls" in the river systems. Many of the major cities and state capitals were established by the colonists at these locations (e.g., Raleigh, North Carolina; Richmond, Virginia; Washington, D.C.; Baltimore, Maryland; Philadelphia, Pennsylvania) because this was as far upstream that the rivers were navigable. This fall line also represents the landward edge of the coastal plain. If all the water tied up in the world's glaciers and ice caps were to melt, then sea levels would rise

Published by Elsevier Science Publishing Company, Inc.
Global Climate Change Linkages
James C. White, Editor

over 225 feet, which would be devastating on a global basis. Fortunately, this prospect is unlikely and could only occur over a time span of millions of years based on the geologic record.

Sea levels have also been much lower in the "recent" geologic past, approaching 300 feet below present levels approximately 15,000 years ago. Resort areas, like Ocean City, Maryland, would have been over 100 miles landward of the shoreline at this time as the seas receded to the shelf edge, exposing the wide continental shelf. The Chesapeake Bay and associated marshes did not exist at this time as the Susquehanna River was rapidly down-cutting its bed. With retreat of the glaciers, the water incorporated in these huge ice sheets, which had extended as far south as northern New Jersey, was released to again start filling up the ocean's basins.

The general effect of sea level rise on coastal lowlands is to induce landward retreat – beaches erode and marshes are lost. Most sandy shorelines worldwide have experienced recession during the past century. Such has also been the case along the U.S. coasts; historical data indicate that about 90 percent of our sandy beaches are eroding. Accelerated sea level rise due to greenhouse-induced global warming will only increase erosion rates and exacerbate the present shoreline dilemma. Also, coastal marshlands will be lost at ever-increasing rates, with the nation's marshes perhaps disappearing at the high rates of loss presently experienced in Louisiana.

Loss of Coastal Wetlands

Wetlands account for most of the land less than 1 m above sea level. These extensive marshes, swamps, and mangrove forests fringe most of the U.S. coastline, particularly along the Atlantic and Gulf coasts. Coastal wetlands serve as nurseries for fish and shrimp, many birds, and fur-bearing animals. They are vital to coastal recreation, to the maintenance of water quality, and as a buffer against shore erosion.

The estimated original extent of salt marshes in the United States was 5 million acres (Hoese, 1967). This acreage has been significantly reduced through a variety of actions, including an early widespread practice of filling marshlands in urban areas. Wetlands loss has also been caused by other human actions, such as the construction of canals and waterways. In response to this loss, federal and state programs have been designed to prevent wetlands destruction. Specifically, Section 10 of the federal Rivers and Harbors Act, Section 404 of the federal Clean Water Act, and Executive Order 11988 on floodplain management all establish permit requirements for actions affecting waterways and wetlands. However, the effectiveness of this permitting process has been questioned; the Congressional Office of Technology Assessment (OTA) concluded that permit applications for wetlands alteration are still rarely denied. The continuing human destruction of wetlands should be kept in mind when considering sea level rise and its potential effects on wetlands deterioration.

Ecological conditions in coastal marshes range from marine to nearly terrestrial. A change in controlling factors, such as water salinity or tidal and wave energy, will cause a displacement in marsh zonation. Generally, coastal marshes are divided into low and high marsh based on their elevation relative to sea level (Redfield, 1972). Since marsh plants are attuned to particular water levels (e.g., *Spartina patens*, salt meadow grass which grows at mean high tide), a rise in sea level will shift the distribution of plant species proportionally landward. Beyond this fundamental response to variation in relative sea level, however, a more complex set of attendant responses may occur, tied to the type of marsh considered. Thus, anticipated changes in coastal marshes must be assessed within the context of the basic marsh types that characterize U.S. coasts.

Marshes have been classified on the basis of the flora present (Redfield, 1972) and salinity and floristic relations (Chabreck, 1972), and functionally on the basis of geologic/geomorphic processes (Stevenson et al., 1986). Nevertheless, with respect to the future effects of a rise in sea level, coastal marshes may be broadly divided into backbarrier marshes, estuarine (brackish) marshes, and tidal freshwater marshes.

Backbarrier Marshes

Backbarrier marshes occur along the bay sides of barrier systems of the Atlantic and Gulf coasts. Studies (e.g., Zaremba and Leatherman, 1986) show that these marshes are formed and destroyed rapidly in such dynamic environments. Maintenance of these marshes is therefore more a function of barrier stability than the pace of upward growth of the marsh surface, since sediment supplies are ample (Letzsch and Frey, 1980). For barriers rapidly migrating landward, there may be a net decline in backbarrier marshes. This has been found to be the case at north Assateague Island, Maryland, where sediment blockage by jetties has greatly increased the rate of landward barrier migration (Leatherman, 1984a), and the same qualitative result would be anticipated as a result of accelerated sea level rise.

Estuarine (Brackish) Marshes

Estuarine marshes embrace a wide variety of floristic species in diverse geologic settings where salinities are less than 30 ppt. These marshes, comprising integral components of major estuarine systems such as the Chesapeake Bay, occur in areas of quiescent waters and ample sediment supply. Accretionary budgets differ widely, but in a dynamic equilibrium condition on-site production of organic materials and influx of mineral sediments cause vertical accretion, balancing the local rate of sea level rise.

Accretion rates generally exceed the local relative rates of sea level rise. Consequently, most estuarine marshes do receive adequate sediment supply to compensate for current sea level rise. This must have been the case over the last few hundred years since "natural" marsh loss has not historically been reported to be a problem. Three notable exceptions occur at Barn Island, Connecticut, Blackwater Marsh, Maryland, and in Louisiana, where the present short-term rates of marsh accretion are lower than the local rates of sea level rise. In Louisiana there is widespread loss of coastal wetlands, in part attributable to a sediment deficit. Marsh deterioration is also occurring in the Blackwater Wildlife Refuge, Maryland, but no such problems are yet reported at Barn Island. Exceptionally low local rates of sediment accretion appear to be the cause in both Connecticut and Maryland.

Tidal Freshwater Marshes

Tidal freshwater marshes are located in the upper reaches of estuaries and other areas where ambient salinities are less than 5 ppt. The flora of these marshes is varied and lacks the typical vegetation zonation of open-coast marshes. The effects of rising sea levels will be saltwater intrusion and the eventual dominance of salt-tolerant plants. However, the effects of canalization on tidal freshwater marshes in the Mississippi delta demonstrate that dramatic increases in salinity over a comparatively short period exceed the capability of these marshes to adjust so that rapid losses ensue.

Processes of Marsh Loss with Sea Level Rise

Land losses in most marshes result from a combination of mechanisms. Shoreline erosion at the seaward edge of the marsh, being the most obvious process, could be expected to accelerate with increased water levels. Nationally, however, shoreline erosion probably accounts for only a small percentage of all marsh losses annually. The comparative resistance of marshy shorelines to wave attack suggests that, with rapidly rising sea levels, most marshes will be long since submerged before extensive shoreline erosion occurs.

A more probable catastrophic mechanism of marsh loss with a large increase in sea levels will be the formation of extensive interior ponds allied with general tidal creek bank erosion and headward growth as tidal prisms increase. The rapid enlargement and coalescence of interior ponds in marshes subject to rapid coastal submergence have been amply documented in the Mississippi delta (DeLaune et al., 1983) and at the Blackwater Wildlife Refuge (Stevenson et al., 1986). The magnitude of marsh losses from interior ponding is instructive. At the Blackwater Wildlife Refuge in Maryland, over one-third of the total marsh area (about 5,000 acres) was lost between 1938 and 1979 by the growth of interior ponds. The physiological mechanism behind the development of interior ponds is believed to be anoxia and ultimate death of marsh plants, as sea levels outpace the ability of the marsh to maintain elevation.

Human-Induced Changes

The most dramatic changes in wetlands have historically resulted from human alterations. Over half the salt marshes in New England have been lost because of dredge and fill activities. Elsewhere, the expanse of marshes has actually been increased by poor land practices. Early settlers felled large tracts of forest for agricultural fields, resulting in massive siltation of some bays and estuaries. This pattern is especially true of the Chesapeake Bay, where the colonial port at Gunpowder River is now separated from navigable waters by several miles of intertidal flats, colonized by marsh grasses.

Present human activities are largely preventing sediments from reaching wetland areas. Moreover, soil conservation practices through contour plowing, buffer strips, and no-till agriculture have substantially reduced the influx of soil into adjacent water bodies and wetlands. Dams and levees on major rivers trap material upstream and prevent over-bank flow of muds and fine sand during flood conditions. This is a particularly acute problem in coastal Louisiana, where the marshes have been established on deltaic sediments and have continued to accrete upward by sediments delivered during flood stage of the Mississippi River. Without these levees and other engineered structures, some cities built on floodplains, such as New Orleans, would be subject to massive and frequent flooding.

Finally, wetlands are being lost in coastal Louisiana because of pipeline and navigation canals that now lace much of the area, allowing saltwater intrusion. The resulting interactions are complicated, but there is no doubt that this practice has significantly contributed to the dramatic loss of wetlands presently being experienced in Louisiana.

Although salt marshes are protected by federal legislation, major losses of estuarine marshes can be anticipated in the future because of bulkheading along bay shores (Figure 1). With Holocene sea level rise, these salt marshes have been naturally translated landward through time. With the construction of landward-flanking bulkheads, which are prevalent along the mainland bay shores of many coastal states, these marshes will literally be squeezed out of existence with sea level rise (Figure 1).

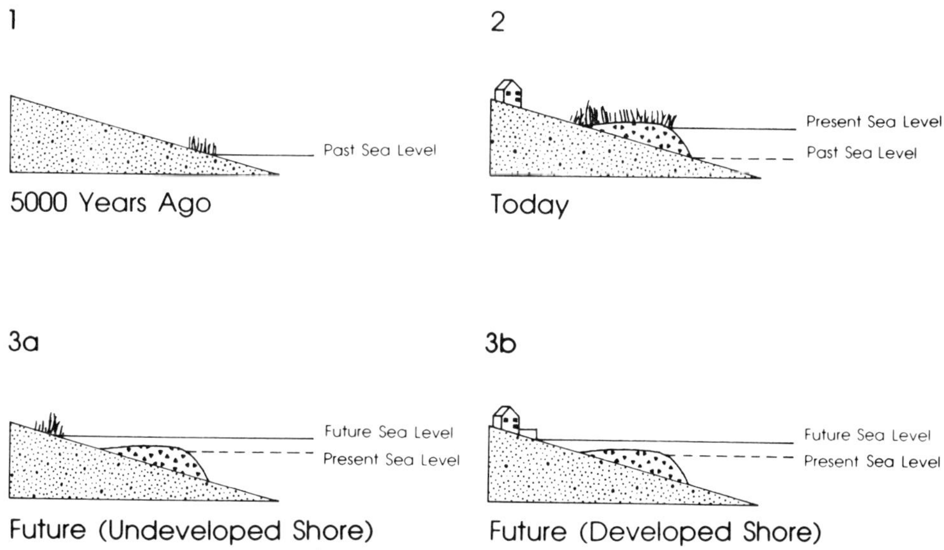

Figure 1. Evolution of marshes as sea level rises.

Prospects for Wetlands

The prospect for wetlands is bleak in light of existing conditions and projected changes. The present situation in coastal Louisiana can be used to forecast qualitatively the expected wetlands changes elsewhere. Due predominately to subsidence and elimination of sediment supply by levee construction, the land surface has been subsiding about 1 cm/yr. Without the influx of massive quantities of inorganic riverborne sediments, the marsh surface can only accrete vertically by biogenic production, which is limited. Therefore, marshes are not able to keep pace with relative sea level rise and are being drowned in place. A rapidly subsiding substrate of accelerating sea level rise can yield similar results. Marsh grasses cannot accrete vertically fast enough to keep pace with sea level rise. This will likely be the fate for extensive estuarine marshes elsewhere in the United States if substantially higher rates of sea level rise are realized in the future.

Beach Erosion

Sea level is one of the principal determinants of shoreline position. There is a causal relationship among several factors: sediment supply, wave energy, sea level, and shoreline position. Rising relative sea level tends to cause shoreline recession, except where this trend is offset by an influx of sediment.

The primary reason that a sea level rise would induce beach erosion is that natural beach profiles are concave upward; this geometry results in the wave energy being dissipated in a smaller water volume than without sea level rise and thus the turbulence generated within the surf zone is greater. The profile responds by conforming to a more gentle nearshore slope which requires additional sediment to be eroded from the beach.

Most sandy shorelines worldwide have retreated during the past century (Bird, 1976). Progradation has been restricted to coastal areas where locally excess sediment is supplied by river sources or where the land is being elevated due to tectonic uplift or isostatic glacial rebound. Human interference cannot be considered a primary cause of erosion worldwide since it is also occurring on sparsely populated and little developed sandy coasts (Bird, 1976). Such recession could result from an increase in storminess, but this trend would have to be almost worldwide to account for erosion on geographically dispersed sandy shorelines. Therefore, in view of the demonstrated general relative rise of sea level along the United States shoreline, the linkage between shore retreat and sea level rise is based on more than circumstantial evidence; it can be stated that the relationship is causal in nature.

In some areas, it is clear that human modifications to the nearshore have caused substantial erosional pressures. Undoubtedly the principal contributor has been the construction of jettied inlets and deepening of channel entrances for navigational purposes. Along shorelines with a high rate of longshore sediment transport, these constructed features trap sediment at the updrift jetty. If material dredged from the navigation channel is not placed on the downdrift beaches, then the amount of downdrift erosion will be equal to the reduction in transport. The present situation downdrift of the Ocean City Inlet jetties, where Assateague Island is being sand-starved and rapidly eroded (Leatherman, 1984a), is indicative of this problem.

Considering the past extreme fluctuations in sea level and shoreline position on a geologic time scale, the phenomenon of erosion occurring today is dwarfed. Yet the historical rise of sea levels of approximately one foot during the past century along the U.S. east coast is very significant when viewed within the context of the building binge (boom) that has occurred along the water's edge in the past few decades. Hundreds of thousands of beachfront structures (exquisite single-family houses, high-rise condominiums, and elegant hotels) have been built within a few hundred feet of an eroding shoreline. In fact, beachfront property is some of the most valuable real estate in the country.

The present dilemma has resulted from this tremendous investment in coastal property at a time when most sandy beaches nationwide, and indeed worldwide, are eroding. Prior to the establishment of permanent buildings on beaches, most people took little note of the ever-changing shorelines. It was popularly envisioned that the beaches erode a little (during winter storms) and then rebuild in the summer. While there is post-storm recovery of beaches (but usually not completely) and most beaches narrow and widen on a seasonal basis, there is a longer-term trend superimposed on these short-term fluctuations.

The effect of rising sea levels is to induce shore retreat in concert with storms, which impart the wave energy needed to transport vast quantities of sediment. A net change of several feet per year is often difficult to measure when the beach width naturally fluctuates by 50 feet or more seasonally, thus confusing the casual observer. However, measured over decades to centuries with accurate means of shoreline mapping (Leatherman, 1983), the trends become apparent and rates of beach erosion can be quantified for particular areas along the coast.

Regional Coastal Evaluation

The U.S. coastline can be divided into major shore reaches based on regional characteristics; this compartmentalization approach allows a general comparison between similar areas. However, erosion rates can be highly variable along a specific stretch of shoreline, and site-specific data are required for precise determinations.

Sea level rise will have a different effect along the U.S. coastline depending upon existing conditions (e.g., geologic history, coastal geomorphology). It is possible to divide the coasts into physiographic regions for consideration of their response to relative sea level rise. For instance, the existing problems in Louisiana do not apply to Maine as the delta region is undergoing pronounced compaction and subsidence while northern New England is experiencing neotectonic uplift.

The present rise in water level is a nested phenomenon, including local, regional and global components. Shoreline position will respond to the cumulative effect of these vertical motions, termed the relative sea level rise, regardless of their cause.

The U.S. continental coastline is highly variable in character but certain regional trends are apparent. Tectonic mapping clearly indicates the reasons for the prominent differences between the Pacific coastal range as compared to the Atlantic and Gulf coastal plains. The tectonically-active Pacific rim is a collision coast where one plate is being subducted below another (Inman and Nordstrom, 1971), resulting in a narrow continental shelf and an essentially nonexistent coastal plain. This coast is characterized by headlands and intervening pocket beaches. By comparison, the Atlantic and Gulf coasts are tectonically stable and constitute trailing edge and marginal sea coasts, respectively.

The Atlantic coastal plain, which extends over 100 miles inland, is characterized by a gently-sloping surface with gradients of only several feet per mile. The prominent landform from Long Island, New York, to Miami Beach, Florida, along the Atlantic coastal plain is barrier islands. The Gulf coastal plain exhibits the lowest average relief and gentlest gradients. Barrier islands are again the dominant coastal landform, but the chain is less continuous than on the Atlantic coast as broken by the Mississippi River deltaic sediments of Louisiana, the marsh outcrops along the "zero-energy" northeast Gulf coast of Florida, and the broad out-flowing of the Everglades along the limestone rocky coast of the southern peninsula of Florida.

These three coasts can be further subdivided into physiographic regions on the basis of geologic history and coastal morphodynamics. Basically eleven types of coasts for the U.S. continent can be defined (Figure 2) using a modification of the classification by Shepard and Wanless (1971).

Atlantic Coastal Plain

The glaciated coast extends from northern New Jersey to Maine. This physiographic region can be further subdivided into the erosion-resistant crystalline (granite) rock of northern New England and the mostly unconsolidated glacial till of southern New England. While coastal barriers have developed by spit growth across many embayments, the mainland is often fairly high near the shore, forming cliffs. The low areas, which are presently subject to storm surges, are clearly marked on topographic maps, and the flooding problem will only increase with accelerated sea level rise.

The coastal compartment barrier chains of New Jersey and Delmarva are characterized by four sectors: (1) terminal north spit, (2) low, eroding headland, (3) long barrier islands backed by open lagoons, and finally to the south, (4) short-stubby barriers with marsh-filled embayments. Within these compartments, there is variable shoreline vulnerability in response to differing rates and patterns of shoreline erosion (Leatherman et al., 1982), storm surge flooding, and inlet breaching potential. Relative sea level rise will have a pronounced effect on this coastline due to its sandy, erodible nature and generally low elevation except for some coastal dune areas. The south shore of Long Island, New York, can also be included in this sector since the glacial outwash plain has a similar physiography to the Atlantic coastal plain.

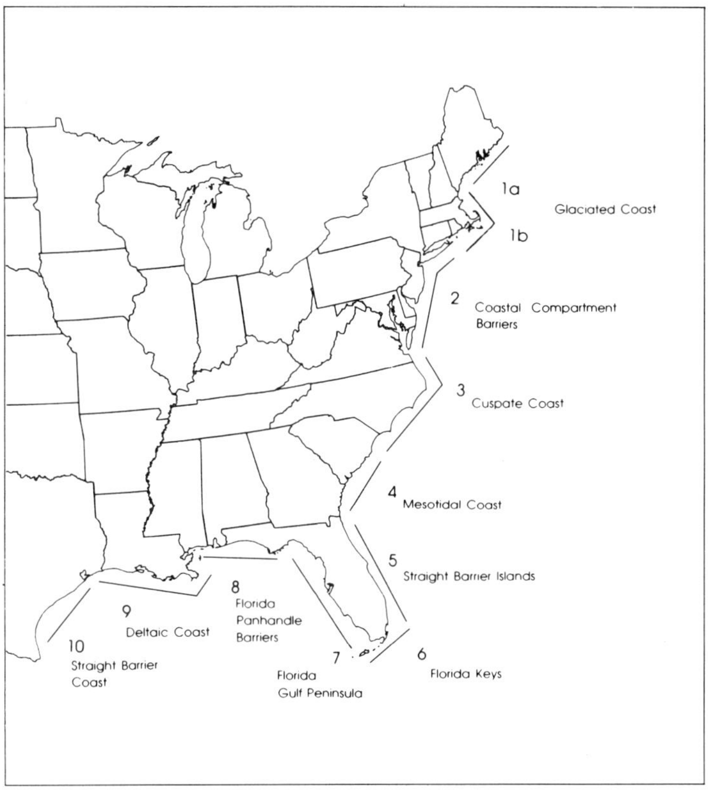

Figure 2. U.S. continental coastal classification based on geologic and morphological criteria.

The cuspate coast, which is best exemplified by Cape Hatteras, North Carolina, extends from Norfolk, Virginia, to Cape Romain, South Carolina. The "Outer Banks" type barrier islands form a string of sand as viewed from space that protrudes far from the mainland coast along the northern part to the point of merger with the mainland at Myrtle Beach, South Carolina. The historical record shows that the Outer Banks of North Carolina have been peppered with inlets (Leatherman, 1988), and sea level rise can result in significant problems along this already eroding, low coast.

The mesotidal coast of the Georgia bight extends from southern South Carolina to northern Florida. This physiographic region has the only tide-dominated barrier islands along the U.S. Atlantic and Gulf coastal plains. These islands tend to be short and stubby with marsh-filled lagoons, often displaying a drumstick shape (Hayes, 1979). The famed Sea Islands (Hoyt and Henry, 1967) are also present in this region; these coastal landforms are distinctive by being composed of a Pleistocene core (often with a land surface above the 500-year flood level). Holocene (modern) sediments are plastered onto the seaward face of the old Pleistocene barrier or separated from it by a small marsh or tidal creek. The Sea Islands and mesotidal barriers are more stable than their microtidal counterparts due to the fact that the tidal inlets are located in pre-Holocene drainage channels and their location shifts between narrowly-defined limits (Oertel, 1979). The fact that these inlets are tidally adjusted to better accommodate storm surge flooding, in combination with the antecedent topography and possibly some present-day riverine sources of coarse sediment, make these islands more stable than those barriers along the rest of the Atlantic and Gulf coasts.

Straight, narrow barrier islands persist from northern Florida to the Florida Keys with Cape Canaveral being the single anomaly. The lagoons are narrow and shallow, and the presence of existing inlets, many of which are stabilized, largely precludes future inlet breaching. The coast is low and sandy with only occasional outcrops of erosion-resistant limestone (Anastasia coquina rock formation) in the beachface.

The Florida Keys are founded on coral reefs, often capped with coral rubble and faced with sandy beaches produced by the hurricane destruction and wave-grinding up of living corals in the surf zone. Coral reef platforms are more resistant to erosion than barrier islands, but their low elevation makes them especially vulnerable to increased flooding and overwash with relative sea level rise. However, coral reefs can continue to grow vertically, which is nature's response to sea level rise, as long as anthropogenic pollution does not foul and kill these living ecosystems. The principal problem of human habitation of the Keys (beyond the shortage of drinking water) is evacuation of thousands of people during hurricanes; accelerated sea level rise will only exacerbate this problem.

Gulf Coast Plain

The Florida Gulf peninsula contains a diverse coast of sandy barrier beaches and swamps. The Everglades empties directly into the Gulf of Mexico along a wide front. This physiographic section is largely controlled by Pleistocene limestone rock which alternately outcrops near sea level to form coastal swamps or lies tens of feet below the present water surface, wherein barrier islands have formed and can persist. The swampy coast, including the zero-energy coast at the northern bound (Tanner, 1960), are little developed, but the intermittent, low elevation, microtidal barrier islands have been highly urbanized near such population centers as Tampa Bay, Florida.

The Florida panhandle barrier island system actually extends to the Louisiana delta off Gulfport, Mississippi. The barriers are largely Holocene in origin, except for a noted Pleistocene core, which constitutes the western, bulbous end of Dauphin Island, Alabama. These beach-ridge dominated barrier islands indicate periods of past geological accretion and strong westwardly migration due to littoral drift.

The deltaic coast of Louisiana is the most vulnerable to relative sea level rise of the entire U.S. continental area. The sediments are largely fine-grained (silts and clays), which are subject to compaction and hence surface subsidence and to erosion along the shoreline. Much of the Louisiana coast, except for the small area of active delta building, is rapidly retreating landward on the order of tens of feet per year (Penland et al., 1985). The natural problem of deltaic substrate compaction is artificially accelerated by the massive withdrawal of fluids (water, oil and gas) from the ground. Hence some areas are sinking at rates of 1 cm per year, resulting in the drowning of salt marshes and pushing the small sandy barriers as thin sand wedges over the adjacent backbarrier sediments. The Chandelieur Islands and Isles Dernieres barrier chains are being fragmented during hurricanes, and it appears that these islands will be lost during the next 100 years even under the present conditions of relative sea level rise (Penland et al., 1985).

The straight barrier coast of Texas has been well studied (McGowen et al., 1977). The small delta of the Brazos River is the only major interruption of these very long, but fairly wide barrier islands. The existing inlets front the mouths of large bays or rivers emptying into the sea (e.g., Brazos and Rio Grande). Otherwise, the great widths (several miles across) of the Texas barriers, microtidal conditions, and shallow lagoons preclude most inlet activity. Unlike most of the Atlantic and Gulf coasts, rivers are still providing some sources of coarse-grained material (sand) to the open-ocean coast so that the barrier islands are not nearly as "sand-starved" as those found elsewhere. Locally, erosion is a problem, such as at Sergeants Beach, and as evidenced by the lack of beach in front of the Galveston seawall.

Pacific Coast

The Pacific coast of California, Oregon and Washington is quite irregular and diverse. There is little to no coastal plain, and cliffs, both resistant hard rock and unconsolidated river-fill sediments, predominate along this tectonically-active Pacific rim. Within small embayments, sandy to gravelly spits can grow, but these landforms are ephemeral geologically speaking and are highly unstable features over the short-term (Komar, 1976). The sheer extent and geomorphic diversity over short distances largely precludes the typing of this coast into natural units and physiographic regions. Therefore, the best guide is to consider the nature of the sediments (erosion-resistant bedrock or unconsolidated sands, gravels and clays) and the topography (high to low cliffs vs. low sandy barrier spits) to define the local areas of present coastal vulnerability and forecast the future problems of accelerated sea level rise.

Discussion and Conclusions

Coastal Wetlands

Coastal wetlands account for much of the land less than five feet above sea level. These extensive marshes, swamps, and mangrove forests fringe most of the U.S. coastline, particularly along the Atlantic and Gulf coasts. These areas are vital to our fisheries industry, maintenance of water quality, and are important in coastal recreation.

Salt marshes exist in a delicate balance with water levels. With gradual sea level rise (due to local subsidence or worldwide changes), marshes can generally keep pace by trapping sediments in the water column and through accumulation of their own organic material (dead stems and leaves). However, an imbalance can develop if sea levels rise significantly faster than deposition on the marsh surface, eventually resulting in waterlogging and loss.

Land losses in most wetlands result from a combination of mechanisms, with shoreline erosion at the seaward edge of the marsh being the most obvious process. This factor can be expected to accelerate with increased water levels, but shoreline erosion probably accounts for only a few percent of all marsh losses annually. Most marshes will be long since submerged before extensive shoreline erosion occurs.

A more probable catastrophic mechanism of marsh loss with a significant increase in sea levels (e.g., several feet) will be formation of extensive interior marsh ponds. These shallow-water bodies enlarge and coalesce at the expense of marsh vegetation in response to rapid coastal submergence. The magnitude of such losses can be quite extensive as shown by studies at the Blackwater Wildlife Refuge in Maryland, where over one-third of the total marsh area (about 5,000 acres) was lost between 1938 and 1979 by this process. These marshes are being lost because sea levels outpace the ability of the marsh to maintain elevation, ultimately resulting in root death to the marsh plants.

Beaches and Coastal Barriers

Most sandy shorelines worldwide have retreated during the past century. Progradation has been restricted to coastal areas where excess sediment is supplied by river sources or where the land is being elevated due to uplift in some northern regions (e.g., Maine). Human interference cannot be considered a primary cause of erosion since retreat also occurs on sparsely populated and little-developed sandy coasts. Such recession could result from an increase in storminess, but this trend would have to be almost worldwide to account for erosion on geographically dispersed sandy shorelines. Therefore, in view of the demonstrated general relative rise of sea level along the U.S. shoreline, the link between shore retreat and sea level rise is based on more than circumstantial evidence; it can be stated that the relationship is causal in nature.

In some areas, it is clear that human actions have caused substantial erosional pressures. Undoubtedly the principal problem has been construction of jettied inlets and deepening of channel entrances for navigation. Along shorelines with high rates of longshore sediment transport, these constructed features trap sediment at the updrift jetty and, if material dredged from the navigation channel is not placed on the downdrift beaches, cause an amount of downdrift erosion equal to the reduction in transport. At some Florida entrances, tens of millions of cubic yards of dredged material have been dumped offshore and thus are permanently removed from the beach. This has resulted in high rates of erosion downdrift of tidal inlet entrances.

Historical data (maps and aerial photographs) can be used to determine rates of shoreline change. Over one hundred years of records are available for the U.S. coasts from the National Ocean Service of the National Oceanic and Atmospheric Administration (NOAA). This accurate informational base has been utilized for large stretches of the coastline (including much of the U.S. Atlantic and parts of the Gulf and Pacific coasts) to determine historical rates of beach erosion and barrier recession.

The national (unweighted) average rate of shoreline recession is slightly more than one foot per year. Along the Atlantic coast, beach erosion has historically amounted to about two to three feet per year, with the Virginia barrier islands exhibiting the highest rates of erosion (tens of feet per year). The Gulf coast states are distinguished by having the highest average erosion rate in the nation (over five feet per year). While parts of the Florida panhandle are fairly stable or only slightly eroding, the deltaic coast of Louisiana is by far the most dynamic (15 feet/year of erosion on average). The Pacific coast is essentially stable on average, considering the fact that more than half of the shore is hard rock. Locally, there are severe erosion problems (e.g., Oceanside, California), and winter storms have caused dramatic loss of very valuable property in recent years where the beaches are already quite narrow (e.g., Malibu during the 1981-82 storm season).

A number of techniques have been developed to project shoreline retreat due to sea level rise (see 1987 National Academy of Sciences report). Accelerated sea level rise due to the greenhouse effect will increase the erosion rate and cause stable beaches to begin eroding. Based upon the projected rates of sea level rise, it is estimated that sandy beaches will erode at two to five times their current rate. This portends major problems for low-lying areas, such as coastal Louisiana, and increased problems at such beach resorts as Ocean City, Maryland, where the historical rate of beach erosion has amounted to two feet per year. Many of the urbanized beaches along the East and Gulf coast barrier chain are already critically narrow and in need of beach nourishment to protect the existing buildings and infrastructure. Such is the case at Ocean City, Maryland, where a \$30 million beach fill project was initiated in 1988. Elsewhere, many other recreational beaches have been nourished (e.g., Atlantic City, New Jersey; Myrtle Beach, South Carolina; Miami Beach, Florida). The expected lifetime of nourished beaches will be greatly foreshortened by accelerated sea level rise, increasing the frequency and magnitude and hence the cost of such activities.

Responses

There are three general responses to accelerated shoreline erosion: (1) retreat from the shore, (2) armor the coast, or (3) nourish the beach. The choice of a response strategy will depend on a number of factors, including socioeconomic and environmental conditions. The decisions reached will likely be site-specific so that each area/community must be evaluated separately.

For highly urbanized areas, such as Miami Beach or Ocean City, the abandonment option is not realistic. The value of such beachfront property with high density and high-rise structures often approaches \$100 million per mile, making beach restoration the most attractive alternative. Elsewhere, seawalls have been constructed to stabilize the shore and protect coastal communities from the ravages of major hurricanes (e.g., Galveston, Texas). For eroding shorelines that are less developed, the decision becomes more difficult. Therefore, the costs and benefits of protection must be weighed against those of retreating from the shoreline.

Acknowledgments

This research has been supported, in part, by the U.S. Environmental Protection Agency. Much of this text was originally written by the author for inclusion in a 1987 National Academy of Sciences report on "Responding to Changes in Sea Level: Engineering Implications."

References

1. Bird, E.C.F., 1976, Shoreline changes during the last century. *Proceedings of the 23rd International Geographical Congress, Moscow*. Elmsford, NY: Pergamon, 54 pp.

2. Chabreck, R.H., 1972, *Vegetation, Water and Soil Characteristics of the Louisiana Coastal Region*. Louisiana Agricultural Experiment Station Bull.

3. DeLaune, R.D., R. H. Baumann and J.G. Gosselink, 1983, Relationships among vertical accretion, coastal submergence, and erosion in a Louisiana Gulf Coast marsh. *J. Sed. Petrol.* 53:147-157.

4. Hayes, M.O., 1979, Barrier island morphology as a function of tidal and wave regime, in S.P. Leatherman (ed.), *Barrier Islands*, Academic Press, Inc., New York, NY, pp. 1-27.

5. Hoese, H.D., 1967, Effect of higher than normal salinities on salt marshes. *Contrib. Mar. Sci.* 12:249-261.

6. Hoyt, J.H. and V.J. Henry, 1967, Influence of island migration on barrier island sedimentation. *Geol. Soc. Amer. Bull.* 78:77-86.

7. Inman, D.L. and C.E. Nordstrom, 1971, On the tectonic and morphologic classification of coasts. *Journal of Geology* 79:1-21.

8. Komar, P.D., 1976, *Beach Processes and Sedimentation*, Prentice-Hall, Englewood Cliffs, NJ, 429 pp.

9. Leatherman, S.P., 1983, Shoreline mapping: a comparison of techniques. *Shore and Beach* 51:28-33.

10. Leatherman, S.P., 1984a., Shoreline evolution of North Assateague Island, Maryland. *Shore and Beach* 52:3-10.

11. Leatherman, S.P., 1984b., Coastal geomorphic responses to sea-level rise: Galveston Bay, Texas. In *Greenhouse Effect and Sea Level Rise*, M.C. Barth and J.G. Titus, eds., New York: Van Nostrand Reinhold, pp. 151-178.

12. Leatherman, S.P., 1988, *Barrier Island Handbook*, Laboratory for Coastal Research, University of Maryland, College Park, MD, 92 pp.

13. Leatherman, S.P., T.E. Rice and V. Goldsmith, 1982, Virginia barrier island configuration: a reappraisal. *Science* 215:285-287.

14. Letzsch, S.W. and R.W. Frey, 1980, Deposition and erosion in a Holocene tidal salt marsh, Sapelo Island, Georgia. *J. Sed. Petrol.* 50:529-542.

15. McGowen, J.H., L.E. Garnev and B.H. Wilkinson, 1977, *The Gulf shoreline of Texas: processes, characteristics, and factors to use*. Texas Bureau of Economic Geology Circ. 77-3, 27 pp.

16. National Research Council, 1987, *Responding to Changes in Sea Level: Engineering Implications*, National Academy of Science Press, Washington, DC, 148 pp.

17. Oertel, G.F., 1979, Barrier island development during Holocene recession, southeastern United States, in S.P. Leatherman (ed.), *Barrier Islands*, Academic Press, New York, NY, pp. 273-290.

18. Penland, S., J.R. Suter and R. Boyd, 1985, Barrier island arcs along abandoned Mississippi River deltas. *Marine Geology* 63:197-233.

19. Redfield, A.C., 1972, Development of a New England salt marsh. *Ecol. Monogr.* 42:201-237.

20. Shepard, F.P. and H.R. Wanless, 1971, *Our Changing Coastlines*, McGraw Hill Book Co., New York, 579 pp.

21. Stevenson, J.C., L.G. Ward and M.S. Kearney, 1986, Vertical accretion in marshes with varying rates of sea level rise. In *Estuarine Variability*, D. Wolf, ed., New York: Academic Press, pp. 241-260

22. Tanner, W.F., 1960, *Florida coast classification*, Trans. Gulf Coast Assoc. Geol. Societies 10:259-266.

23. Zaremba, R.E. and S.P. Leatherman, 1986, Vegetative physiographic analysis of a U.S. Northeast barrier system. *Environ. Geol. Water Sci.* 8:193-207.

THE LINKAGES BETWEEN CLIMATE CHANGE AND ACID RAIN

Hans C. Martin

LRTAP Office
Atmospheric Environment Service
Environment Canada
4905 Dufferin Street
Downsview, Ontario M3H 5T4

Introduction

I am uneasy about this topic. Having taken part not only in the scientific debate relating to acid rain, but also in the negotiations for controls with Canadian industry and, bilaterally, with the United States, I suffer from a paranoia when the acid rain issue risks being placed on a long-term agenda. Rightly or wrongly, I am cautious about linking acid rain with climate change because it then runs the risks of falling into such a long-term agenda. As we develop stronger linkages there is an implication that the scientific aspects of the issues must be addressed together, the solutions must be evolved together and the schedule for correction must be adjusted to allow climate change to "catch up."

This approach is not acceptable in Canada. The Canadian Government and this population at large consider acid rain a serious national issue and stringent, costly action has been taken to correct the situation. Canada has an SO_2 control program in place which will reduce sulphur dioxide emissions by 50% by the year 1994, based on 1980 levels. It is generally not known that this program, which was initiated a few years ago, will cost approximately $500 million/year over a twenty-year period.

Our discussion today asks the question, "Does a linkage exist?" If yes, then in Canada we tacitly ask, "Are we making a mistake to control sulphur dioxide without a complete understanding of the linkages?" There are three possible outcomes from our current domestic policy to control sulphur dioxide and nitrogen oxides, viz:

A) The control program will reduce the anticipated magnitude of climate change, that is warming.
B) The control program will have no impact on climate change.
C) The control program will aggravate climate warming.

Of course, the worst scenario would be that the actions we are taking now to control sulphur dioxide would aggravate the climate warming predicted for future decades,

Copyright 1989 by Elsevier Science Publishing Company, Inc.
Global Climate Change Linkages
James C. White, Editor

Option C. In this instance it would be better for Canada to do nothing. Following this line of reasoning, the Canadian Acid Rain Control Program would provide a premature solution and more research would be required.

I believe this assessment is naive and perhaps sinister. It uses the increasing complexity of anthropogenic alterations of the chemistry of the environment as a justification for seeking further scientific explanations and clarification before the alteration process is stopped and, perhaps, reversed. Given our limited understanding of the relationships among the growing list of pollutants which are changing the atmosphere and the chemical and biological consequences, the no-action approach will commit us indefinitely to an unknown future with unforeseeable damage. The status quo will prevail until all relationships are known perhaps even beyond climate change, ozone depletion and acid rain, and until the implications of all combinations of possible controls are calculated in the absence of uncertainty. In other words, we cannot move until we know it all. In what follows, I will show that we can address the acidification problem now, with overall beneficial results on the global atmospheric pollution problem.

Before examining the two issues of acid rain and climate change, I would like to clarify my definition of linkage. A linkage is a relationship between two processes or phenomena where a change in one process (e.g. acidification) will affect the other process (e.g. climate warming), and/or vice versa. My definition does not restrict itself to the simple co-existence of the two processes; rather, it takes into account the interactive nature of the processes at work.

The Linkages

In the following sections I will review what we understand about the nature of acidification and climate warming. From the outset it should be noted that there is one fundamental difference between the two issues, namely, the climate warming scenario is based on prediction, based on observations derived from models while the acidification scenario is based on past and present observations as well as models predicting future states. The acidification issue has a sounder scientific base. There is a body of evidence which suggests that climate warming has already started. I do not consider that body of information compelling enough to change my view of the predictive nature of the climate warming scenario.

In what follows, the characteristics of the two issues are compared and linkages are highlighted where they occur. The examples of linkage are not classified according to relative importance except for the qualification "weak," in some cases. The omission might be considered a weakness though I am not sure a rigorous ranking is even possible.

The Gases

The gases involved in acidification are sulphur dioxide and nitrogen oxides. These gases undergo oxidation processes in the atmosphere to produce sulphuric and nitric acids. Nitrogen oxides also are involved in the production of tropospheric ozone, a process which is legitimately included in the discussion of acid rain (more properly called the long-range transport of air pollutants). Nitrogen oxides and volatile organic compounds under appropriate conditions produce tropospheric ozone, notably in warm, humid, summer conditions.

Linkage—Any increase [decrease] in nitrogen oxides, and associated increases [decreases] in tropospheric ozone, will lead to higher [lower] concentrations of the greenhouse gas ozone and, hence, will increase [reduce] climate warming.

The greenhouse gases are different from the primary acidifying gases. They include carbon dioxide, chlorofluorocarbons, nitrous oxide, methane and tropospheric ozone. In addition, the main greenhouse gas, CO_2, existed in the pristine atmosphere (at about 70% of the current level) and along with water vapour was responsible for the earth's "habitable" temperature. In contrast, the acidification gases SO_2 and NO_x have negligible concentrations in the pristine atmosphere and very high concentrations (of the order of x 10 above pristine levels) in those regions where acidification is taking place.

Another difference between the gases is their concentration variability. The high concentration of acidifying gases is a regional phenomena while for CO_2, at least, concentrations are rather uniform globally with no identified hot spots.

Sources

The acidification gases are produced by the transport sector (largest contributor to nitrogen oxides), the nonferrous smelting industry (largest contributor of sulphur dioxide emissions in Canada) and the thermal power industry (the largest contributor of sulphur dioxide in the United States and a large contributor of nitrogen oxides).

Carbon dioxide is produced by all burning processes including those which produce acidification gases. In addition, chemical production (i.e. chlorofluorocarbons), deforestation (the loss of a sink for CO_2), agricultural practices, and natural processes (e.g. production of methane in wetlands) all contribute to the climate warming process.

Timeframe

The acid rain issue has been with us for several decades. Acidification has extensively altered the chemical and biological character of ecosystems and is expected to continue to alter the environment in the future. Corrective action is being undertaken by many countries including the Canadian SO_2 Control Program noted earlier.

The implementation of measures to control climate warming is a matter for the future. Our scientific understanding and predictive capabilities must continue to evolve over several decades. Corrective action to halt or reverse future warming will not take place for some time, perhaps for 5-10 years.

Space Scale

The acidification process is a regional phenomenon largely confined to eastern North America, and central and northern Europe. Future acidification areas are expected to include parts of Africa, China, and South America. These are regions where industrial expansion and the burning of fossil fuels are expected to increase dramatically.

The climate warming phenomenon is global. However, the predicted change in air temperature is nonuniform, with the greatest impacts occurring in the temperate and polar regions. Other changes such as rainfall amounts and global circulation patterns will vary with location.

Atmospheric Residence

The acidification gases cause damage to the environment after they have left the atmosphere. In contrast, for the greenhouse gases, their presence in the atmosphere is the problem.

There are three processes which can occur during the atmospheric lifetime of gases: *transformation, transport* and *deposition.*

Primary acidification gases undergo *transformation* in the atmosphere. Sulphur dioxide oxidizes to sulphate, nitrogen oxides become nitrate and, finally, nitrogen oxides in the presence of volatile organic compounds and appropriate physical conditions produce ozone. In all of these cases the transformation of the primary gases to secondary products is enhanced in a warm atmosphere.

Linkage – The heating of the atmosphere as a result of climate warming will increase the transformation rates of primary acidifying gases and the production of ozone. This change in transformation rate will lead to a change in the relative amounts of acidifying materials deposited and in their concentrations.

Increased production of ozone in a warmed atmosphere has an interesting implication. Since ozone is a greenhouse gas, the increased production of ozone will enhance warming as noted earlier (under "Gases") and thereby further increase the production of ozone. This might be called a loop rather than a link. The loop, or positive feedback, is weak, however, since the limiting factor for the production of ozone in the summer is likely to be the availability of nitrogen oxides and volatile organic compounds, rather than temperature per se. It is possible to identify other loops.

The *transport* of acidifying gases through the atmosphere determines, among other things, the regional patterns of *deposition* and of concentration. With the advent of atmospheric warming due to climate change, it is predicted that the general flow patterns and the precipitation patterns will change, and overall tropospheric instability will increase.

Linkage – The change in general flow will alter the transport trajectories of primary and secondary acidifying materials and, therefore, alter the subsequent deposition patterns of these gases and their regional concentrations.

Linkage – The increased instability in the warmer atmosphere will increase the deposition rate of acidifying pollutants.

Linkage – The predicted changes in precipitation patterns associated with climate warming will lead to a change in the ratio of wet to dry deposition of acidifying pollutants.

These linkages between acidification and climate warming may have public policy implications.

Effects

The general properties of acidifying gases and greenhouse gases are different in a profound way: when acidifying gases fall to the surface they interfere with the chemical and biological stability of ecosystems, whereas, the greenhouse gases change the physical characteristics of the atmosphere and subsequently alter the atmospheric energy balance in situ leading to increases in air temperature.

The acidification of lakes in North America has been one of the most important consequences of the acidification process. With the advent of a warming climate, it is anticipated that lakes in general will become more productive.

Weak Linkage – Higher productivity in lakes is expected to lead to an enhanced capability for biological material to reduce acidity with the result that climate warming will tend to counteract part of the acidification process. Due to the weak nature of this link, there is a risk that the final result of the two processes may be warmer acidified lakes.

Although the relationship between acidification and forest health has not been satisfactorily clarified in North America, it is generally accepted that, among other things, acidification leads to a disturbance of the nutrient status of forest soils with the leaching of trace nutrients as well as the mobilization of some toxic materials. On the other hand, ozone is a known toxin causing injury to trees.

Weak Linkage – Climate warming is expected to improve forest growth if we ignore the anticipated increase in insects and fire. In this way the climate warming can improve the forest's tolerance of the stresses including those due to acidification. This linkage is speculative and does not account for nutrient deficiencies which would still exist in an acidifying regime enjoying an improved climate.

Linkage – Ozone will be produced in larger amounts from nitrogen and volatile organic compounds in a warmer climate.

Weak Linkage – The current forest decline in eastern North America, due to a number of factors including air pollution and acidifying materials, will lead to a reduction of the CO_2 sink provided by growing forests and, hence, will aggravate the warming process. This linkage is expected to be small in temperate zones where the current acidification process is most evident but where the sink is not nearly as important as in the tropics.

The predicted warming will place higher demands on power because of the increased need for summertime air conditioning.

Linkage – The summertime productions of sulphur and nitrogen compounds from power plants will increase, leading to increased acidification and ozone production at the worst time of the year, the time when plants and animals are most susceptible to injury.

The foregoing is intended to be a sampling of linkages which are most apparent, not a complete inventory. There are additional relationships and linkages that might be postulated including some relating to agriculture, human health and materials. The further development of such a list can become academic and speculative, and tends to confuse and to add unnecessary uncertainty.

Correct Now, Correct Later

Acidification is a current issue requiring the correction of an existing problem over large regions of the Northern Hemisphere where extensive damage is being observed and where the effects can be corrected through known and available measures. By contrast the climate warming issue concerns a future state of the environment which is predicted by models, among other techniques, and where a solution is not yet clear and where the acceptance and implementation of solutions will likely require enormous courage and determination.

It is fair to say that the climate change models available today are an acceptable basis for initial policy decisions which will address the warming phenomena. However, policy makers and the public at large are not yet prepared to undertake such policy decisions. This caution is consistent with the state of scientific knowledge. Our understanding of the pace and magnitude of the change is limited as is our grasp of the consequence, the effects. The issue requires further maturation. In fact, the exchange of ideas occurring today at this conference is part of the maturation.

In contrast acid rain solutions are available today and are being applied. In the following paragraphs, I will review some of the measures that have been undertaken in Canada and indicate some of the benefits and disbenefits which are expected for both the acidification and the warming issues:

Smelters: Industrial process changes in the past and anticipated over the next years will dramatically reduce the emissions of sulphur dioxide from nonferrous smelters and reduce carbon dioxide emissions. Both issues will benefit.

Catalytic Converters: Catalytic converters have dramatically cut the emissions of nitrogen oxides from the transport sector. The controls have reduced the potential production of nitric acid in the atmosphere and tropospheric ozone. The net result is the reduction of both acidification and warming. Both issues benefit. There is a small energy penalty of about 1% associated with catalytic converters.

Coal Washing: Coal washing reduces sulphur dioxide emissions and improves the efficiency of thermal power plants reducing CO_2 emissions. Hence, coal washing benefits both the acid rain and the climate warming issue.

Low NO_x burners: In power plants this technology reduces nitrogen oxides emissions and, hence, tropospheric ozone production. In this way this technology benefits both the acidification problem and the climate warming problem.

Scrubbers: These devices, installed in power plants, reduce sulphur dioxide emissions. There is an energy penalty of up to 10% associated with scrubbers. This penalty necessitates the additional burning of coal and, therefore, increased carbon dioxide production. Hence, scrubbers benefit the acid rain issue but aggravate to some degree the climate warming problem.

One could go on with further examples. The conclusion drawn here is that the application of scrubbers to thermal power plants represents the only disbenefit which might be associated with technologies which are now used to address the acid rain issue. To put this disbenefit in the proper perspective, it is worthwhile to examine the Canadian situation more closely.

In Canada, coal combustion produces 20% of the domestic CO_2 emissions. If we assume all of this coal is utilized in power plants, then the 10% energy penalty will lead to a 2% increase in carbon dioxide emissions (10% of the 20%). This increase in Canadian carbon dioxide emissions should be balanced against the benefit of a major reduction in the SO_2 emissions.

There has been no mention of alternate approaches to the problem of carbon dioxide, nitrogen oxides, and sulphur dioxide emissions. The alternate approaches in clude conservation and improved efficiency in the use of power. Both measures lead to reduced emissions and benefit both issues.

In North America and Europe the acidification issue has been a training ground for scientists, policy makers, politicians and the general public. We have:

- Organized cross-disciplinary research;
- Developed ecosystem/regional approaches;
- Trained scientific staff;
- Synthesized data from research on sources, atmospheric transport/transformation, effects, socio-economic consequences, etc.;

- Developed solutions/options;
- Applied solutions;
- Evaluated solutions.

We have not yet finished the job. The problem will not go away. The problem cannot be avoided by focusing on another problem which is far larger, poorly defined, and in the future.

Conclusion

We are conducting massive, unpredictable experiments with our atmosphere. We have messed up the laboratory for whatever reason. Our survival depends upon a thorough clean-up, a determined, expensive, step-wise clean-up which continues for decades.

There is a public beyond these walls which is relying on groups like ours here today to evaluate, make judgments and shepherd solutions. There is no future in hand wringing and moaning over acid rain plus climate change plus ozone depletion plus toxics, etc. It is time to treat the good news that we have with objectivity and to disseminate it to an anxious public. The news is not that scientists are piling more complex, perplexing problems on top of each other year by year, with no solutions in hand. The good news is that something can be done and is being done to correct one of our pervasive environmental problems – the acidification process. The solutions being applied are not perfect but they are good, and they are contributing to corrections that we will have to make to deal with climate warming. That is a positive linkage.

LINKAGES BETWEEN CLIMATE CHANGE AND STRATOSPHERIC OZONE DEPLETION

Robert C. Worrest

U.S. Environmental Protection Agency Stratospheric Ozone Research Program
U.S. E.P.A. (RD-682)
401 M Street SW
Washington, DC 20460

Katie D. Smythe and Alexander M. Tait

Science and Policy Associates,
Washington, DC 20005

Introduction

Alterations of the earth's atmosphere by human activities are now of regional and global proportion. In the last decade scientists, policy makers, and the general public have focused increasing concern on ozone depletion in the upper atmosphere and on global climate change. These two problems are closely interrelated and both are of truly global scale. Two primary areas link the issue of stratospheric ozone depletion to global climate change: atmospheric processes and ecological processes.

Atmospheric processes establish a linkage through the dual roles of certain trace gases in promoting global warming and in depleting the ozone layer. The primary radiatively active trace gases are carbon dioxide, nitrous oxide, chlorofluorocarbons, methane, and tropospheric ozone. In the troposphere, the atmosphere up to 10 miles above the earth's surface, these compounds function as greenhouse gases. At increased levels they can contribute to global climate change. Many of these gases also influence the concentration of ozone in the stratosphere, the atmospheric layer located between 10-30 miles above the earth's surface. This diffuse layer of ozone in the stratosphere protects life on earth from harmful solar radiation. A reduction of this layer could have very important impacts on the earth's systems.

The second mode of interaction revolves around various ecological processes. Physical, chemical, and biological activities of plants and animals are affected directly by global climate change and by increased ultraviolet radiation resulting from depletion of stratospheric ozone.

Copyright 1989 by Elsevier Science Publishing Company, Inc.
Global Climate Change Linkages
James C. White, Editor

The purposes of this paper are to: 1) provide general background on the stratospheric ozone depletion issue, and 2) discuss the linkages surrounding both the atmospheric and ecological processes.

Background

The natural distribution of ozone in the earth's atmosphere is not uniform. It is concentrated primarily in a thin layer in the stratosphere where it blocks most of the ultraviolet radiation in the 290 to 320 nm range from reaching the earth's surface. This is known as the ultraviolet-B (or UV-B) range and it can be damaging to humans, biological organisms, and man-made materials.

Many gases emitted from man's industrial and agricultural activities can accumulate in the earth's atmosphere and ultimately contribute to alterations in the vertical distribution of stratospheric ozone. These gases accumulate in the lower atmosphere (troposphere) and then gradually migrate upward into the stratosphere where they contribute to depletion of stratospheric ozone. The atmospheric and chemical processes involved in the destruction of ozone are extremely complex and are reviewed elsewhere [e.g., 1], but the following section will summarize one of the most important interactions.

Scientists have demonstrated that in recent years the release into the atmosphere of halogens containing chlorine and bromine – such as chlorofluorocarbons (CFCs) and halons – has resulted in the destruction of the protective ozone layer in the stratosphere. Short wavelength radiation hitting the stratosphere causes a breakup of the chlorofluorocarbons releasing the chlorine radicals (see Figure 1). A chlorine radical destroys stratospheric ozone through a catalytic cycle producing oxygen molecules. After converting ozone to oxygen, the radical – now chlorine oxide – then reacts with an oxygen atom from another ozone molecule and emerges unchanged, ready to destroy more ozone.

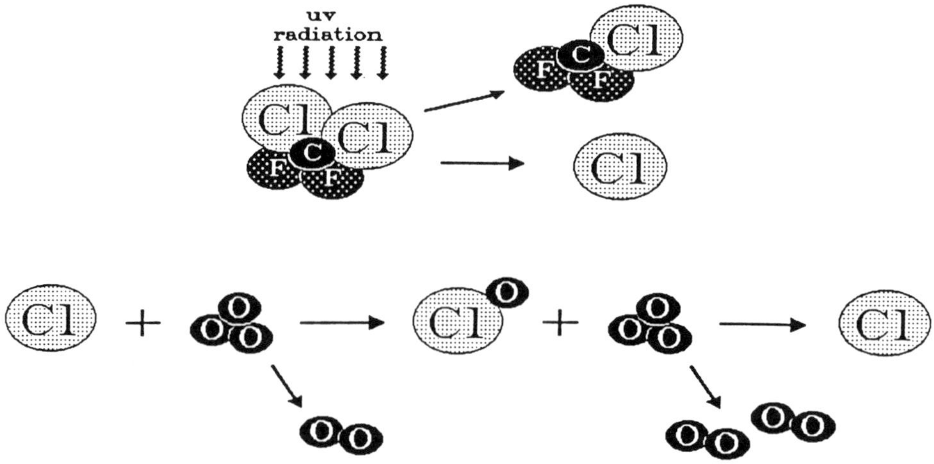

Figure 1. The catalytic cycle of a chlorine radical breaking down ozone into oxygen molecules.

Less ozone in the stratosphere will result in a greater transmission of ultraviolet radiation to the surface of the earth, causing detrimental effects. Because of the long atmospheric lifetimes of CFCs and halons, scientists expect stratospheric ozone to continue to decrease into the middle of the next century even if emissions are curtailed worldwide [2].

TABLE I. Trace gases affecting ozone concentrations and global climate change.

Trace Gas (Formula)	Primary Source	Ave. Life in Atmos.	ODP*	GP**
CFC-11 ($CFCl_3$)	Refrigerant/AC, Plastic Foams, Aerosols	75 yrs	1.0	0.40
CFC-12 (CF_2Cl_2)	Refrigerant/AC, Plastic Foams, Sterilants	110	1.0	1.00
CFC-113 ($C_2F_3Cl_3$)	Solvents	90	0.8	0.3-0.8
Halon 1211 (CF_2ClBr)	Fire Exting.	25	3.0	?
Halon 1301 (CF_3Br)	Fire Exting.	110	10.0	0.80
Carbon Tetrachloride (CCl_4)	Industrial Processes	67	1.1	0.05
Methyl Chloroform (CH_3CCl_3)	Industrial and Natural Processes	8	0.1	0.01
Nitrous Oxide (N_2O)	Fossil Fuels	150	--	0.016
Methane (CH_4)	Biogenic Activity, Fossil Fuels	11	--	0.001
Carbon Dioxide (CO_2)	Fossil Fuels	7	--	0.00005
Carbon Monox. (CO)	Motor Vehicles	0.4	--	--

* ozone depletion potential (CFC-11 = 1.0)

** greenhouse potential (CFC-12 = 1.0)

Table 1 provides a summary of the major trace gases that contribute to ozone depletion, including their primary uses, atmospheric lifetime, and projected annual increase. Currently, the primary focus of regulations and mitigation options are on CFCs and halons, as depicted in the Montreal Protocol [3]. The treaty document was officially ratified in December 1988; regulatory provisions become effective in July 1989. The agreement calls for a freeze of CFC production (at 1986 levels) by 1989, a 20 percent decrease in production by 1993, and an additional 30 percent decrease by 1998.

Several recent national and international work group reports detailed discussions of the growing concern about stratospheric ozone depletion and assessment of scientific bases underlying such concern [e.g., 1, 4, 5]. A report by the Ozone Trends Panel has also highlighted the issue of stratospheric ozone depletion [6]. The panel's report describes the global decrease in stratospheric ozone in recent years and the probable role of CFCs and other ozone-depleting compounds in the development of the Antarctic ozone hole.

Atmospheric Interactions

Many of the atmospheric trace gases, such as chlorofluorocarbons, are relatively transparent in the visible region of the solar spectrum. They do, however, absorb the long wavelength radiation that is radiated back from the surface of the earth, resulting in the greenhouse effect. In addition to CFCs, other gases contribute to the greenhouse effect, including: carbon dioxide, methane, nitrous oxide, and tropospheric ozone. The effects of these gases on global warming is cumulative because each blocks different wavelength radiation. Their concentrations have been increasing over time, some more dramatically than others. This point is particularly important because of their relatively long atmospheric lifetimes, especially CFCs, halons and nitrous oxide.

Chlorofluorocarbons (CFCs)

The release of CFCs and other chlorine-containing compounds decreased initially in the 1970s as a result of regulatory action to ban selected, non-essential CFC compounds used as aerosol propellants. This regulation was enacted in several countries, including the United States. Currently, CFC production and consumption are increasing due at least in part to the involvement of newly industrialized and lesser developed countries in CFC use.

In addition to breaking down ozone through the action of chlorine radicals, chlorofluorocarbons contribute to the greenhouse effect. Estimates put the CFC contribution to global temperature change at 20-25 percent, although under the Montreal Protocol it would be reduced to 15-20 percent.

Nitrous Oxide (N_2O)

Nitrous oxide in the atmosphere originates from both natural and man-made sources, including many bacterial processes involved in the nitrification or denitrification cycles. Recently, nitrous oxide has been increasing at a rate of about 0.2% annually. In the atmosphere N_2O is partly converted into nitrogen oxides (NO_x).

Nitrous oxide (N_2O) plays a role in both ozone depletion and global climate change. The N_2O functions as a greenhouse gas contributing to global warming. Converted to nitrogen oxides (NO_x) it destroys stratospheric ozone in a catalytic cycle similar to that of chlorine radicals. On the other hand, nitrogen oxides can serve as a temporary sink for the ozone depleting chlorine monoxide (ClO), so the net effect is uncertain. NO_x also is a precursor to acid deposition.

Carbon Dioxide (CO_2)

The concentration of CO_2 has been increasing in recent years by an average of about 0.5% annually. Carbon dioxide links the issue of stratospheric ozone depletion to that of the global climate change issue primarily because of its role as a greenhouse gas. As such, it will absorb solar radiation being radiated from the surface of the earth and re-radiate it in all directions increasing the global warming. Because it is a greenhouse gas CO_2 modifies the temperature structure of the atmosphere, cooling the stratosphere. Less ozone is destroyed if the stratosphere is cooler, so the effect of CO_2 acting alone is to decrease ozone depletion in the stratosphere.

Methane (CH_4)

Methane has been increasing fairly constantly at a rate of about 1% annually. Resulting from both natural and man-made processes, methane is involved in several important reactions in the atmosphere. Through its effect on the amounts of water vapor in the stratosphere methane can lead to destruction of stratospheric ozone. Water droplets may act as a surface upon which the reactions that destroy ozone occur. The increase or decrease of ozone depletion will depend upon where the water vapor is produced. Ozone depletion will increase if water vapor increases in the stratosphere. But an increase in water vapor in the troposphere will heighten the greenhouse effect—water vapor is the most important greenhouse gas—and decrease ozone depletion through cooling the stratosphere.

A further effect of methane occurs in the troposphere. Here, when methane is oxidized, it will produce an increase in the amount of tropospheric ozone. This reaction occurs in the presence of nitrogen oxide.

Natural sources of methane include natural wetlands, arctic tundra, agricultural crops such as rice paddies, and ruminate animals. Man-made sources of CH_4 include the production of fossil fuels such as natural gas and oil, and cement production. The natural processes contribute about half of the total methane production.

Carbon Monoxide (CO)

Carbon monoxide is not a radiatively important trace gas, but it is involved indirectly in both stratospheric ozone destruction and global warming. Carbon monoxide controls the concentration of the hydroxyl radical (OH) in the troposphere, which has a direct effect on the concentration of methane. The concentration of methane, as described earlier, plays a role in the amount of tropospheric ozone as well as stratospheric ozone. Methane is also a very important contributor to the greenhouse effect. Although calculations are highly variable, estimates indicate that the concentration of carbon monoxide is increasing at about 1% per year.

Stratospheric and Tropospheric Ozone (O_3)

Based on current scenarios, the stratospheric component of the total ozone column is calculated to decrease over time, whereas the tropospheric constituents of the total ozone column will increase. Whether or not there will be an overall increase in the level of ultraviolet radiation reaching the surface of the earth is still uncertain. Even if the levels of ultraviolet radiation at the earth's surface were not to change, the photochemical reactions involving ultraviolet radiation and the interaction of ozone with various other atmospheric gases would alter the distribution of ozone in the atmospheric column. Changes in the concentration and altitude of ozone will play a major role in altering temperature and atmospheric processes affecting current climate and perhaps add to long-term global climate change.

Increased levels of ultraviolet radiation reaching the earth's surface also will increase the production of ozone at ground level through photochemical reactions. These conditions affect regional air quality. Tropospheric ozone formation takes place in the presence of nitrogen oxide. In addition, hydrogen peroxide (H_2O_2) is produced, which is a strong oxidant and a catalyst in the production of sulfuric acid from sulfur dioxide. These two processes illustrate the linkage between stratospheric ozone depletion and acid deposition. Sulfur dioxide and nitrogen oxides are the two major precursors to acid deposition.

Trace gases affecting ozone also contribute to global climate change. Any efforts by humans to address the potential problems in either area will influence the other. If global warming were to begin, efforts to address the rise in greenhouse gases could increase ozone depletion. Restraints imposed on the buildup of carbon dioxide, methane, and nitrous oxide to control their contribution to global warming, might reduce their role as moderators of potential ozone depletion in high CFC emission scenarios [7].

Ecological Interactions

A major consequence of decreasing the ozone layer is an increase in the transmission of UV-B radiation to the earth's surface. Scientists have identified many potentially serious effects on the environment and on human health from increased exposure to UV-B radiation. These include damage to: agricultural crops, forests, marine organisms, human health (eye disease, immune system, skin cancer), and certain materials. Changes associated with an altered global climate, such as increased CO_2 levels, interact with the effects of UV-B radiation.

Terrestrial Ecosystems

In assessing the impact of increased exposure of crops and terrestrial ecosystems to UV-B radiation it must be recognized that existing knowledge is in many ways deficient. The effects of enhanced levels of UV-B radiation have been studied in species from only a few representatives of the major terrestrial ecosystems. We derive most of our knowledge from studies focused upon agricultural crops and conducted at mid-latitudes. Despite uncertainties due to the complexities of field experiments, the data presently available suggest that plant photosynthesis is vulnerable to increased levels of solar UV-B radiation [1]. Unlike drought or other geographically isolated stresses, stratospheric ozone depletion would affect all areas of the world, including ecosystems whose UV-B sensitivity has not been investigated.

In general, UV radiation causes reduced leaf and stem growth, lower total dry weight, and lower photosynthetic activity in sensitive cultivars of plant species [8]. Less photosynthesis decreases the amount of CO_2 fixed by plants and exacerbates the rise in CO_2 levels. Higher CO_2 may lead to global warming. This shows a direct link from an effect of stratospheric ozone depletion through terrestrial ecosystems to climate change.

Increased levels of UV-B radiation also may affect forest productivity. Only limited data are available on coniferous species, but about one-half of the species of seedlings studied were adversely affected by UV-B radiation [9]. Existing data also suggest that increased UV-B radiation will modify the distribution and abundance of plants. Even small changes in competitive balance over a period of time can result in large changes in community structure and composition [10].

Aquatic Ecosystems

Current evidence indicates that ambient solar UV-B radiation is an important limiting ecological factor in marine ecosystems. Even small increases of UV-B exposure could result in significant ecosystem changes [11]. In marine plant communities a change in species composition rather than a decrease in net production would be the probable result of enhanced UV-B exposure [12]. A change in community composition at the base of food webs may produce instabilities within ecosystems that likely would affect higher trophic levels [13].

Inhibition of marine microbial activity by increasing UV-B radiation could have important consequences for several global biogeochemical cycles. Phytoplankton photosynthesis in the upper layer of the oceans provides a sink for approximately 80% of the anthropogenic CO_2 released to the atmosphere. Bacterial activity in the oceans provides probably the most important global source of CH_3I and CH_3Cl (the only significant natural source of chlorine to the stratosphere). Microorganisms in aquatic ecosystems produce large quantities of methane and nitrous oxide. Chemical and photochemical oxidation of natural organic matter in water bodies produces carbon monoxide. Enhanced UV-B radiation, resulting from stratospheric ozone depletion, can alter these processes and affect the levels of the various greenhouse gases in the atmosphere.

Human Health

Depletion of the ozone layer, resulting in large part from CFCs in the stratosphere, leads to increased UV-B radiation. The UV-B radiation has many adverse effects on human health. Kripke [14], van der Leun [15], and the U.S. EPA [1] summarize the potential health effects of UV-B radiation. Probably the best defined health effects are increases in skin cancer cases expected to result from even small increases in UV-B radiation reaching the earth's surface. However, skin cancer affects only a small minority of the world's population, light-skinned Caucasians, and therefore is not a significant global problem.

Another important health concern—with potentially more extensive impacts on more diverse human populations than skin cancer—is the effect of increased UV-B radiation on human immune suppression. Adverse ocular effects have also been documented from enhanced UV-B radiation, including an increase in the incidence of cataracts in exposed populations.

CFCs, as mentioned above, are important greenhouse gases and contribute to global climate change. Several elements of climate change could have important human health effects. Higher temperatures will increase heat stress in humans, especially among the elderly. Increased temperature combined with changes in precipitation distribution could alter the geographic distribution of diseases and parasites, providing vectors for the northward movement of tropical diseases.

State of Knowledge

Table 2 illustrates differences in our state of knowledge regarding various anticipated biological effects and their potential global impact, as viewed by an expert subcommittee of EPA's Science Advisory Board [16]. One of the present dilemmas we face is that our current state of knowledge of effects that have the greatest potential for widespread global impacts is low. For example, the current knowledge of potential effects of increased UV-B radiation on the human immune system is relatively low, but the global impact on human health could be quite high.

Policy Information Needs

Many uncertainties remain regarding the effects of stratospheric ozone depletion and global climate change. With both of these issues, scientists should structure research around questions that are explicitly relevant to policy decisions. Using "policy-relevant questions" as the framework for research provides the rationale for undertaking specific initiatives. The questions that follow are based on fundamental information needs relevant to stratospheric ozone depletion, including a question addressing the information needed about linkages between ozone depletion and climate change (question 4).

1) How widespread are the potential effects of UV-B radiation? What populations, systems or substances are at greatest risk? An increase in UV-B radiation will continue to exert some degree of influence on man and the environment for the foreseeable future. This stress will transcend regional and national boundaries, exerting significant ef-

TABLE II. Potential effects of increased UV-B radiation resulting from decreased stratospheric ozone

Effects	State of Knowledge	Potential Global Impact
Plant Life	Low	High
Aquatic Life	Low	High
Skin Cancer	Moderate to high	Moderate
Immune System	Low	High
Cataracts	Moderate	Low
Climate Impacts*	Moderate	Moderate
Tropospheric Ozone	Moderate	Low**

* Contribution of both stratospheric ozone depletion itself and gases causing such depletion to climate changes.

** Impact could be high in selected areas typified by local or regional scale surface-level ozone pollution problems.

Modified from Kripke [16].

fects on widely separated ecosystems and populations. These two attributes make understanding the effects of UV-B radiation on biological and human systems of critical importance.

2) How intensive are the effects of UV-B irradiance? What are the doses that would pose a significant risk? The response of systems to UV-B radiation may differ significantly for different life stages of the same species as well as by latitude and season. What are the specific dose-response relationships, and what mechanisms are involved? Identification of basic and common mechanisms of damage will allow extrapolation of dose-response models to other systems not yet studied.

3) If the dose of UV-B radiation is reduced, what is the anticipated rate and extent of recovery of sensitive systems? Organisms and ecosystems generally (but not always) recover from environmental stress. It is important to know to what degree and at what rate individual organisms and ecosystems can repair the damage inflicted by enhanced UV-B irradiance. In managed ecosystems (agriculture and silvaculture), UV-B tolerant species may replace less tolerant species. In natural ecosystems some species may be able to adapt over time to increasing UV-B irradiance.

4) What interactions occur between increased UV-B radiation, global climate change and atmospheric pollutants? The composition of the atmosphere depends to a large degree on the natural and man-made emissions of greenhouse gases such as carb-

on dioxide, CFCs, methane, and nitrous oxide. If new data suggested that increasing UV-B radiation would alter biogeochemical cycles involving these chemicals and result in a disruption of ecosystems or change in global climate, additional regulation to protect the stratospheric ozone layer may be necessary.

5) What mitigation options can be applied? To which systems? Chemicals with less ozone-depleting potential can be used as alternative substances, such as for aerosol propellants. But substitutes are not currently available for some ozone-depleting compounds. Recovery and reuse of substances also is a viable option. We also need to develop mitigative responses to the effects of UV-B radiation, including the replacement of sensitive species with more UV-B tolerant species. Options relevant to the human system are more tenuous and are important to investigate.

Policy makers and scientists should periodically review and revise this list of questions to reflect the current state of science.

Current Activities

The depletion of stratospheric ozone and subsequent increase in UV-B radiation is a problem of global proportions, and one that is not occurring in an environmental vacuum. Other changes are occurring, including global climate change and acid deposition/air pollution. Scientists must examine these issues collectively through an integrated research approach.

A concerted effort is recently underway in the United States and numerous other countries in response to plans and initiatives under the International Geosphere-Biosphere Program (IGBP), also known as the Global Change Program. IGBP, which was launched in 1986 by the International Council of Scientific Unions [17], is the first international attempt to integrate research on global climate change and stratospheric ozone depletion.

Reports by the National Academy of Sciences' Committee on Global Change [18], and the White House's Committee on Earth Sciences [19] describe the work proposed by the U.S. federal agencies. Agencies involved in the U.S. Global Change Program include the following:

Department of Agriculture (USDA)
Department of Energy (DOE)
Department of the Interior (DOI)
Environmental Protection Agency (EPA)
National Oceanic and Atmospheric Administration (NOAA)
National Science Foundation (NSF)
National Aeronautics and Space Administration (NASA)

Success in addressing the global problems facing us today relies upon such international and interagency coordination. It is also important to address the related global problems as parts of a whole; stratospheric ozone depletion, global climate change, and acid deposition are intricately linked.

References

1. U.S. EPA. 1987. Risks to crops and terrestrial ecosystems from enhanced UV-B radiation. Pages (11)1-31 In: *Assessing the Risks of Trace Gases that Can Modify the Stratosphere*, J. Hoffman (ed.) USEPA 400/1-87/001C. U.S. Env. Prot. Agency, Washington, DC.

2. Crawford, M. 1987. Landmark ozone treaty negotiated. *Science*, **237**, 1557.

3. United Nations Environment Programme. 1987. *Montreal Protocol on Substances that Deplete the Ozone Layer* -Final Act. UNEP, Nairobi, Kenya.

4. Schneider, T.; Lee, S.D.; Grant, L.D.; & Wolters, G.; eds. 1988. *Atmospheric ozone research and its policy implications: third US-Dutch international symposium*. Elsevier, Amsterdam, The Netherlands (in press).

5. World Meteorology Organization (WMO) and Canada Department of the Environment. 1988. *Proceedings of the World Conference on the Changing Atmosphere.*

6. Watson, R.T. and Ozone Trends Panel; Prather, M.J. and Ad Hoc Theory Panel; and Kurylo M.J. and NASA Panel for Data Evaluation. 1988. *Present State of Knowledge of the Upper Atmosphere 1988: An Assessment Report*. NASA Reference Publication 1208. National Aeronautics and Space Administration, Office of Space Science and Applications, Washington, DC.

7. Miller, A.S. and I. Mintzer. 1986. *The sky* is *the limit: strategies for protecting the ozone layer*. Research Report #3, World Resources Institute, Washington, DC.

8. Tevini, M. and W. Iwanzik. 1986. Effects of UV-B radiation on growth and development of cucumber seedlings. In: *Stratospheric Ozone Reduction, Solar UV Radiation and Plant Life*, R.C. Worrest and M.M. Caldwell (eds.), Springer-Verlag, Heidelberg.

9. Teramura, A.H. and J. Sullivan. 1988. *Annual Report to the U.S. Environmental Protection Agency: The Effects of Changing Climate and Stratospheric Ozone Modification on Plants*. Univ. of Maryland, College Park, Maryland (USA).

10. Gold, W.G. and M.M. Caldwell. 1983. The effects of ultraviolet-B radiation on plant competition in terrestrial ecosystems. *Physiol. Plant.* **58**, 435-444.

11. Damkaer, D.M. 1982. Possible influence of solar UV radiation in the evolution of marine zooplankton. In: *The Role of Solar Ultraviolet Radiation in Marine Ecosystems*, J. Calkins (ed.), Plenum, New York.

12. Worrest, R.C. 1983. Impact of solar ultraviolet-B (290-320 nm) upon marine microalgae. *Physiol. Plant.* **58**, 428-434.

13. Kelly, J.R. 1986. How might enhanced levels of solar UV-B radiation affect marine ecosystems? In: *Effects of Changes in Stratospheric Ozone and Global Climate*, J.G. Titus (ed.). U.S. Environ. Prot. Agency and United Nations Environment Prog., Washington, D.C.

14. Kripke, M. 1988. Health effects of stratospheric ozone depletion: an overview. In: Schneider, T.; Lee, S.D.; Grant, L.D.; & Wolters, G. (eds.) *Atmospheric ozone research and its policy implications: third US-Dutch international symposium.* Elsevier, Amsterdam, The Netherlands (in press).

15. van der Leun, J.C. 1988. Effects of increased UV-B on human health. In: Schneider, T.; Lee, S.D.; Grant, L.D.; & Wolters, G. (eds.) *Atmospheric ozone research and its policy implications: third US-Dutch international symposium.* Elsevier, Amsterdam, The Netherlands (in press).

16. Kripke, M. 1987. *Review of EPA's Assessment of the Risks of Stratospheric Modification.* Prep. by the Stratospheric Ozone Subcommittee, Science Advisory Board. SAB-EC-87-025, U.S. Environ. Prot. Agency, Washington, DC.

17. International Council of Scientific Unions. 1986. *The International Geosphere-Biosphere Program: A Study of Global Change.* Rept. No. 1. ICSU 21st General Assembly, Bern, Switzerland.

18. National Academy of Sciences. 1988. *Toward an Understanding of Global Change: Initial Priorities for U.S. Contributions to the International Geosphere-Biosphere Program.* National Academy Press, Washington, DC.

19. Committee on Earth Sciences. 1989. *Our Changing Planet: A U.S. Strategy for Global Change Research.* Rept. to Accompany the President's FY90 Budget. Washington, DC.

LINKAGES BETWEEN GLOBAL CLIMATE WARMING AND AMBIENT AIR QUALITY

Perry J. Samson, Stephen J. Augustine, and M. Sanford Sillman

Department of Atmospheric, Oceanic, and Space Sciences
University of Michigan, Ann Arbor, Michigan 48109-2143

Abstract

With the possibility of global warming due to increasing concentrations of carbon dioxide, methane, chlorofluorocarbons and other greenhouse gases, it behooves us to evaluate the possible ramifications of atmospheric warming on ambient air quality problems. This paper examines possible influences of warming on urban and regional ozone levels and regional visibility degradation. Potential linkages between atmospheric warming and tropospheric ozone production are outlined. These linkages suggest that urban and regional ozone problems and regional visibility may worsen after a global warming. There is evidence that the high temperatures of the summers of 1987 and 1988 helped produce an increase in the number of occurrences of ozone in exceedence of the National Ambient Air Quality Standard. Similar evidence is being studied for increases in occurrences of reported haze. Extrapolation of these findings to the possible warmer conditions in the industrial Midwest and northeastern United States suggests that current ozone standards may be even more difficult to achieve than previously thought.

Introduction

As the concentrations of greenhouse gases continue to rise, there is increasing speculation that atmospheric temperatures will rise on a global level [1,2]. In this paper we assume that temperatures will rise globally and that this will lead to increases in temperatures in the United States, or at least increases in the occurrence of high temperature episodes. From that assumption we develop a number of potential linkages between atmospheric warming and ambient concentrations of ozone (O_3) and sulfate ($SO_4^=$). Finally we show that evidence already exists for these linkages and that these linkages could make it very difficult for the United States to achieve the National Ambient Air Quality Standards (NAAQS) for ozone in the near future.

The list of gases in the atmosphere ranked in roughly their order of availability is presented in Table 1. Of these gases nitrogen and oxygen account for about 99% of all

Copyright 1989 by Elsevier Science Publishing Company, Inc.
Global Climate Change Linkages
James C. White, Editor

molecules in the atmosphere. Those gases which absorb radiation in the "long-wave" section of the radiation spectrum[1], and hence contribute to atmospheric warming, are identified along with the gases considered to be "criteria pollutants" by the United States Environmental Protection Agency. Ozone stands out as being both a greenhouse gas and a criteria pollutant.

TABLE 1. Ranked order of tropospheric gases with those most important to atmospheric warming denoted by * and the criteria pollutants denoted by †.

Nitrogen (78%)	Xenon
Oxygen (21%)	Ozone*†
Water*	Ammonia
Argon	Ethane
Carbon Dioxide*	Methyl Chloride
Neon	Carbonyl Sulfide
Helium	CFC-12*
Methane*	CFC-11*
Krypton	Formaldehyde
Hydrogen	Sulfur Dioxide†
Nitrous Oxide*	Nitrogen Oxides†
Carbon Monoxide†	Other CFCs*

Ozone is also unique in that it is both advantageous and detrimental to human health. Figure 1 shows the structure of the atmosphere near the Earth's surface. The troposphere is the lowest layer of the atmosphere extending up to about 10 km from the surface. This layer contains about 80% of the mass of the atmosphere (and hence about 80% of all greenhouse gases). High ozone in this layer is detrimental to human health and the United States Environmental Protection Agency has set a NAAQS of 0.12 ppm not to be exceeded in any hour of a day more than three times in any sequential three-year period.

[1]The spectrum of electromagnetic radiation is divided into "short-wave" radiation, which includes the wavelengths of transmission from the sun (less than about 1.5 μm), and "long-wave" radiation, which includes the wavelengths of transmission from the earth (greater than about 1.5 μm).

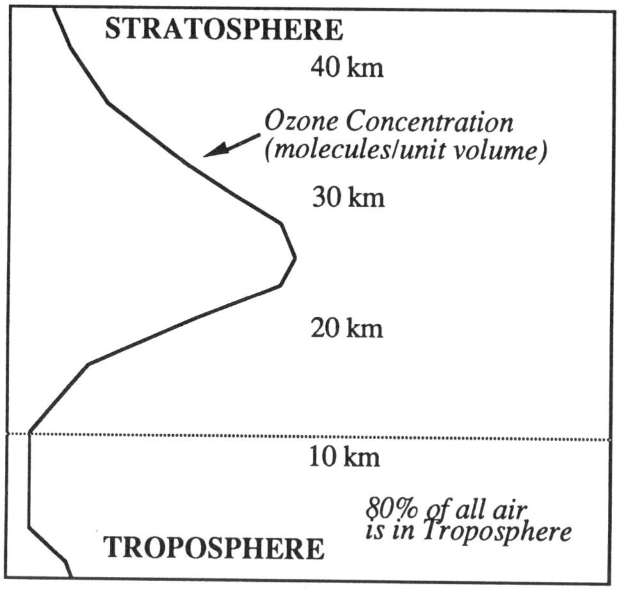

FIG. 1. Schematic of ozone distribution within troposphere and stratosphere. Given that about 80% of the mass of air is located in the troposphere, the majority of atmospheric warming will occur there. While the ozone peak in the stratosphere protects the earth from ultraviolet radiation, the secondary maximum near the ground can be detrimental to humans.

Potential Linkages for Chlorofluorocarbons

The family of chemicals known as chlorofluorocarbons (CFCs) are extremely efficient absorbers of long-wave radiation. Figure 2 illustrates that CFCs are thought to be the second most important greenhouse gas at this time despite their relatively low concentration (generally parts per trillion as opposed to parts per million of carbon dioxide and methane) in the atmosphere.

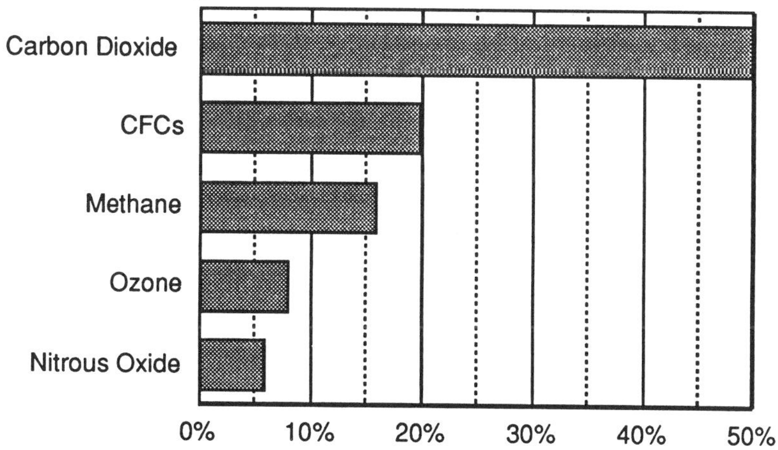

FIG 2. The percentage of tropospheric warming attributable to five greenhouse gases in the 1980s (from Hansen et al. [3]).

The potential link of CFCs to ambient air quality come from two processes. First, the CFCs are known to be at least partially responsible for depletion of ozone in the stratosphere. This process, as illustrated in Figure 3, could allow increased amounts of ultraviolet radiation to reach the troposphere. Additional radiation could potentially increase the production of ozone in the troposphere. The magnitude of ozone rise attributable to this effect has been estimated by Penner et al. [4] to be of the order of 10% assuming a 25% reduction in stratospheric O_3. Second, the CFCs in the troposphere act as a greenhouse gas and could enhance tropospheric warming directly in concert with the enhanced warming due to increases in tropospheric ozone. The warming could then feed back into the production of ozone via the mechanisms discussed next.

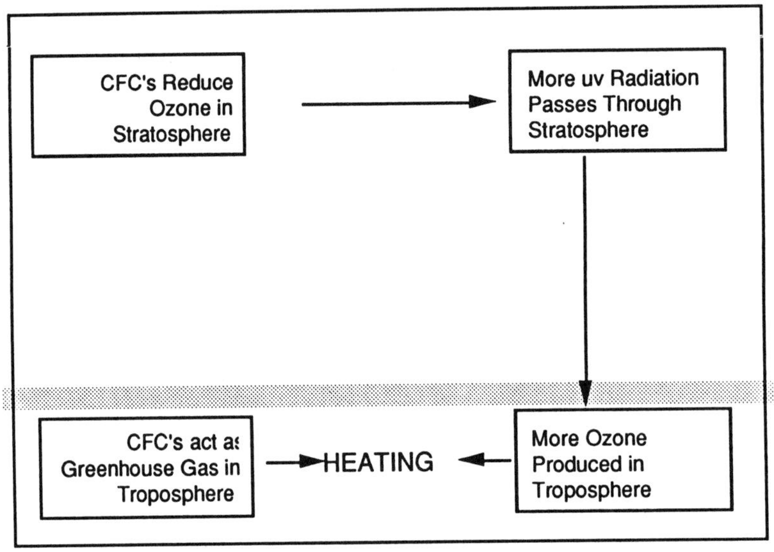

FIG 3. Schematic of the feedback CFCs could play in the production of tropospheric ozone and tropospheric warming.

Potential Linkages for Ambient Ozone

Ozone (O_3) is produced in the troposphere in the presence of sunlight as a result of the photolysis of nitrogen dioxide (NO_2). When nitrogen oxide (NO) and NO_2 are present in sunlight, NO_2 is destroyed by the reaction

$$NO_2 + \text{sunlight} \rightarrow NO + O \tag{1}$$

where O is atomic oxygen. The atomic oxygen will react with available oxygen molecules (O_2) in the presence of other molecules, M, as

$$O + O_2 + M \rightarrow O_3 + M \tag{2}$$

NO_2 is formed primarily by reaction of emitted NO with O_3:

$$O_3 + NO \rightarrow NO_2 + O_2 \tag{3}$$

Reaction (3) followed by (1) and (2) result in no net change in O_3. However, an additional source of NO_2 occurs through the reaction of NO with VOCs. Conversion of NO to NO_2 via reactions with VOCs, followed by reactions (1) and (2) cause a net increase in O_3. Thus the production of O_3 is intimately tied to the availability of NO_x (= NO + NO_2) and VOCs (see Seinfeld.[5] for a more complete overview of photochemical mechanisms).

The potential linkages between tropospheric warming and the increased production of tropospheric ozone are illustrated in Figure 4. With rising temperatures (or increased occurrence of high temperature days) there exists a higher potential for:

- increased emissions of volatile organic compounds, notably isoprene and alpha-pinene, from biogenic sources;
- increased emissions of VOCs from anthropogenic sources;
- increased photochemical production due to a change in reaction rates; and
- potentially increased occurrence of the meteorological conditions associated with high ozone levels.

The figure also indicates that, if increases occur in the level of tropospheric ozone, there may be an increase in heating due to the increased absorption of long-wave radiation by the atmosphere.

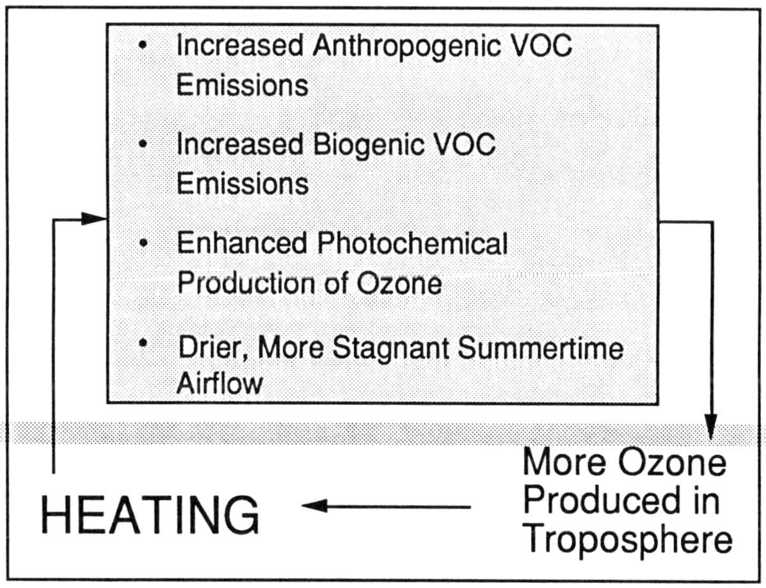

FIG 4. Mechanisms whereby tropospheric heating could cause the enhancement of tropospheric ozone production. The increased ozone could then further exacerbate the tropospheric heating as a greenhouse gas.

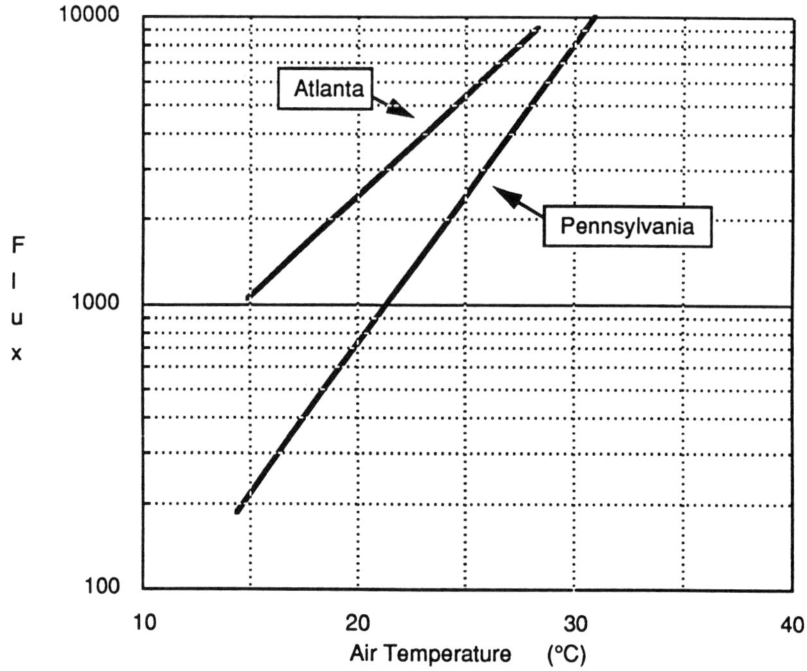

FIG 5. Variations in flux of isoprene from deciduous forests reported by Lamb et al. [6].

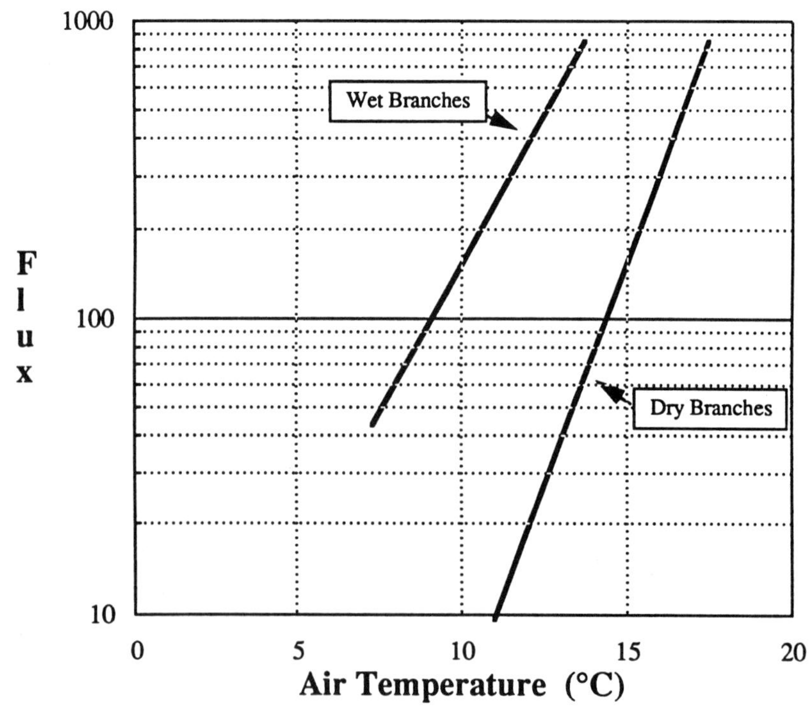

FIG 6. Variations in flux of alpha-pinene from coniferous forests reported by Lamb et al. [6].

Biogenic emissions of VOCs are believed to increase with rising temperature. Isoprene emissions from deciduous forests have been shown to increase exponentially with temperature. Figure 5, illustrates the exponential growth observed by Lamb et al. [6].

Likewise, biogenic emissions of alpha-pinene from coniferous forests have been shown to increase exponentially with temperature. Figure 6 shows the exponential increases in alpha-pinene flux from a coniferous forest with temperature as reported by Lamb et al. [6].

Anthropogenic Emissions of VOCs are now believed to be increasing with temperature. In newer automobiles evaporative losses of VOCs increase with temperature, in part because of the design of fuel-injected vehicles which return fuel vapor to the gas tank. It is believed that these systems may be venting VOCs at higher temperatures because of a combination of the design of the vapor return system and the recent increases in the vapor pressure of gasoline. The increase in the vapor pressure of gasoline has increased the vapor return system pressures to a point where the system pressure can exceed design levels and be released into the atmosphere through a pressure release valve. It is believed that at high temperatures (generally greater than about 30°C) fuel-injected automobiles may be emitting VOCs at rates 10 times previously thought.

Other anthropogenic sources probably increase with temperature. The volatilization of VOCs is known to increase with temperature so it would be reasonable to assume that evaporative emissions from a variety of petrochemical sources may increase with temperature. Also, the increased demand for electrical power to supply peak demand during heat waves could lead to increased emissions from power generation facilities.

The Chemical Mechanisms known to lead to ozone production are influenced by temperature. A recent evaluation by Penner et al. [4] suggested that in an urban setting the rate of increase of ozone production with temperature could be substantial. Her results, reproduced in Figure 7, also hold when applied to a rural situation. Applying the model of Sillman et al. [7] to scenarios of changing temperature and of changing temperature and the changing emissions of isoprene with temperature we find that increases in ozone production are proportionally more modest than for the urban situation, but nonetheless significant. The main reason for the increased production of ozone is the increased rate of thermal decomposition of peroxyacetyl nitrate (PAN) at higher temperature [8].

Meteorological Conditions may also change with increasing tropospheric temperature. With temperatures there is often a reduced occurrence of large temperature contrasts across the eastern half of the United States. Thus fewer synoptic storm systems will occur and, in general, wind speeds will be reduced. Lower wind speeds would allow for more pollutant buildup, low cloud cover and precipitation, increase photolysis and decrease removal processes.

While climate model estimates of the degree of temperature rise have generally been consistent between models, there is virtually no agreement on the changes in precipita-

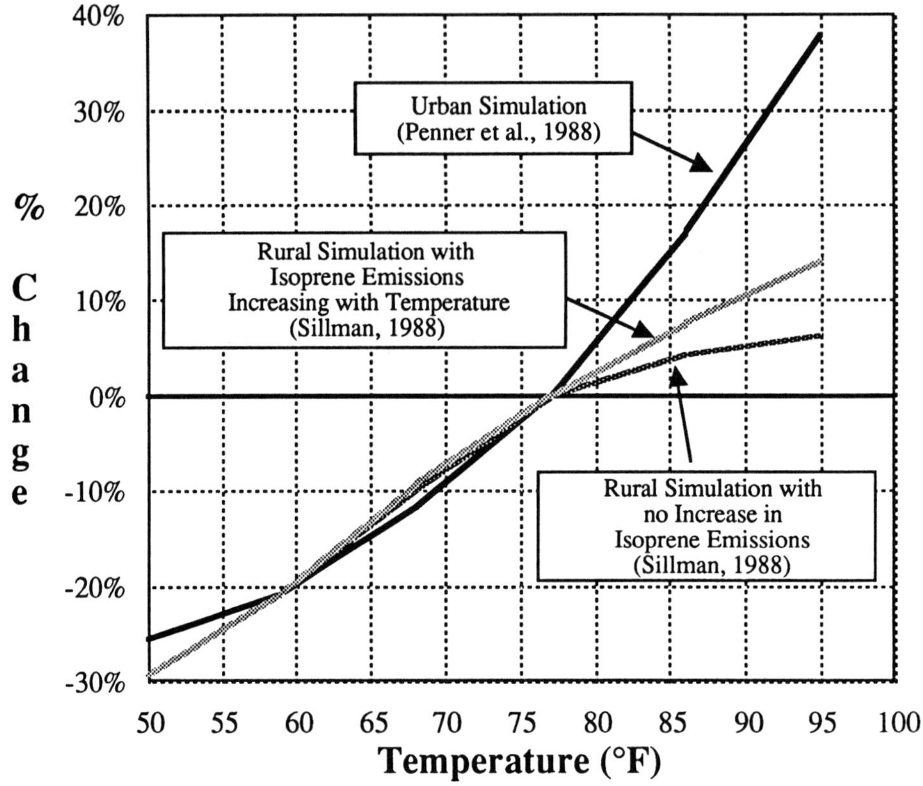

FIG 7. The impact of changing temperature on production of tropospheric ozone. Urban impacts were developed by Penner et al. [4]. Rural sensitivity was developed using the model of Sillman et al. [7].

tion amounts or atmospheric circulation patterns over scales the size of eastern North America [9]. Thus the probability of atmospheric circulation changing with tropospheric warming is speculative but not unreasonable.

Potential Linkages for Ambient Sulfate

The potential effects of tropospheric warming on ambient sulfate ($SO_4^{=}$) is less well-defined. On one hand it is easy to argue that $SO_4^{=}$ levels would increase because of the increased emission of its precursor, sulfur dioxide (SO_2). Additionally the increased availability of VOCs would help speed the homogeneous chemical transformation of SO_2 to $SO_4^{=}$. On the other hand, the increased $SO_4^{=}$ could lead to reductions in visibility (since the particle size of $SO_4^{=}$ will efficiently scatter visible wavelengths). With increases in tropospheric sulfate more sunlight will be attenuated.

Evidence of Linkages

Evidence already exists linking higher temperatures with high ozone days. Figure 8 shows that a strong relationship exists between maximum daily temperature and maximum ozone concentrations in Ann Arbor, Michigan.

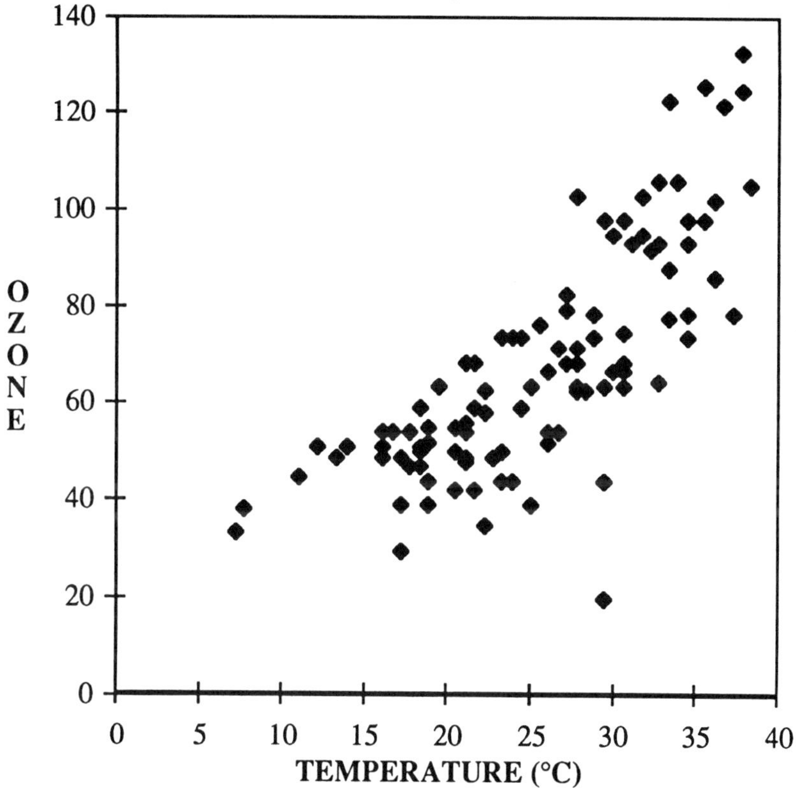

FIG 8. Maximum ozone concentration (ppb) in Ann Arbor during the period April to October, 1987 and 1988 as a function of maximum daily temperature.

Changes in Meteorological Conditions

It is hypothesized that meteorological conditions may change if global temperature rise were to occur. A method for stratifying days into periods of similar synoptic air flow has been developed. The method is based upon the stratification of three-day periods into categories of similar wind flow at 850 mb (roughly 1500 meters above mean sea level). The stratification has been used to identify which meteorological situations have the highest likelihood of producing sulfur deposition at multiple locations across eastern North America [10].

This research has identified a set of 19 meteorological categories based on similarity in wind flow at 850 mb. Of these categories we have found that three of the categories have been most often associated with high ozone days. Figure 9 shows the three-day

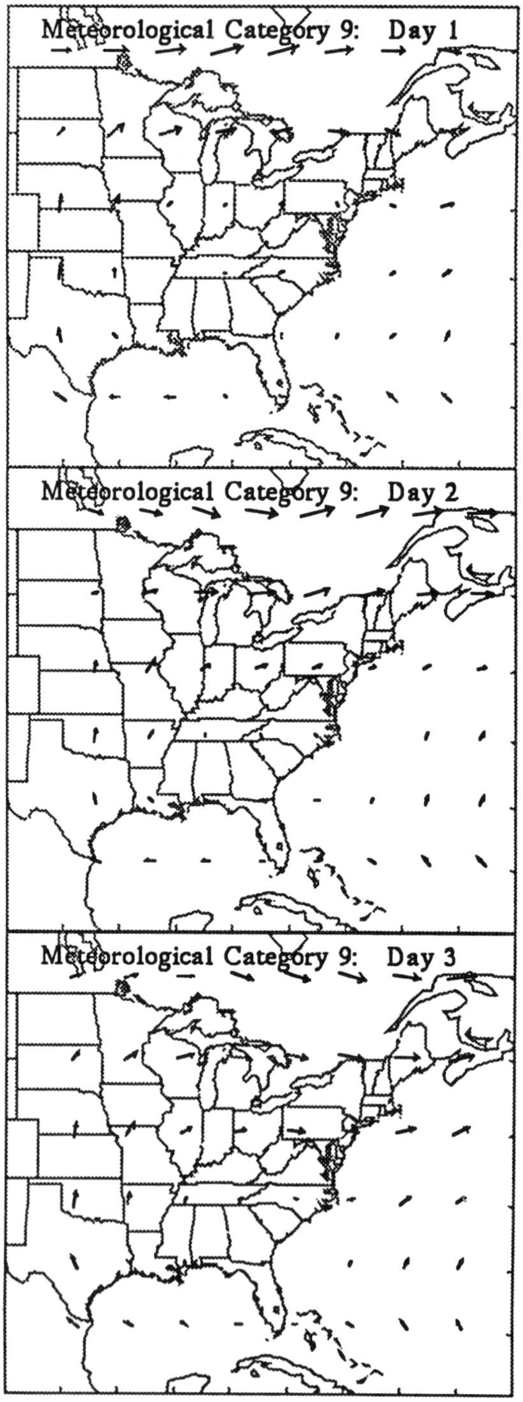

FIG 9. Three day sequence of air flow at 850 mb (about 1500 m above mean sea level) for one of the meteorological categories most often associated with high ozone concentrations. The length of the wind vectors is proportional to the wind speed.

progression of air flow characteristic of one of the three categories (arbitrarily numbered 9).

The pattern shown in Figure 9 has relatively low wind speeds, which can enhance the buildup of ozone and ozone precursor concentrations. The pattern is also often associated with above average temperatures and is most likely to occur during the summer period. Two other flow categories (arbitrarily numbered 7 and 8) which occur predominantly in the summer also were found to have a large number of high ozone associated with them. They also contain relatively low wind speeds with anticyclonic motion over eastern North America.

Figure 10 illustrates that the occurrence of categories 7, 8 and 9 is associated with higher than average temperatures in Ann Arbor, Michigan. The figure shows the temporal trends in mean May maximum temperatures as a function of year and the frequency of categories 7, 8 and 9 occurrence. While these results are exploratory, it suggests that warmer summers are more likely to see slower moving weather patterns.

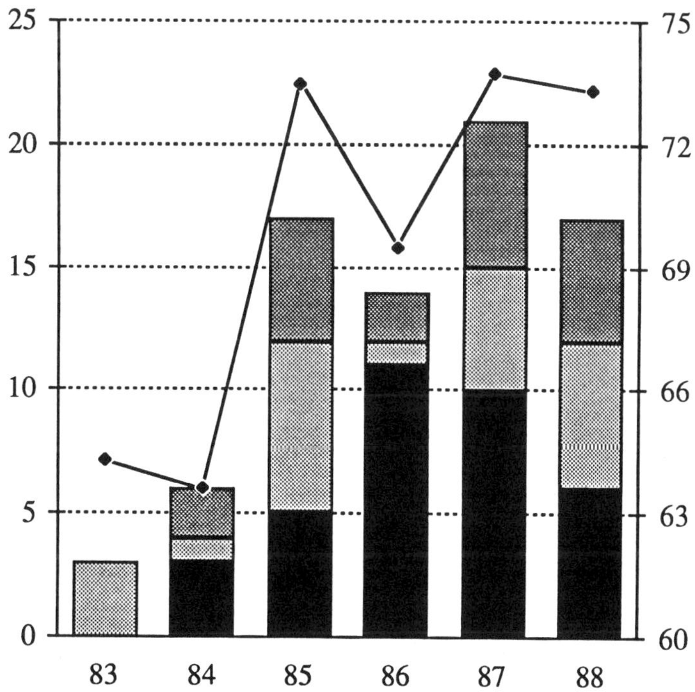

FIG 10. The mean maximum temperature in May in Ann Arbor, Michigan, as a function of the frequency of occurrence of three meteorological categories known to produce above average temperatures and ozone over northeastern United States.

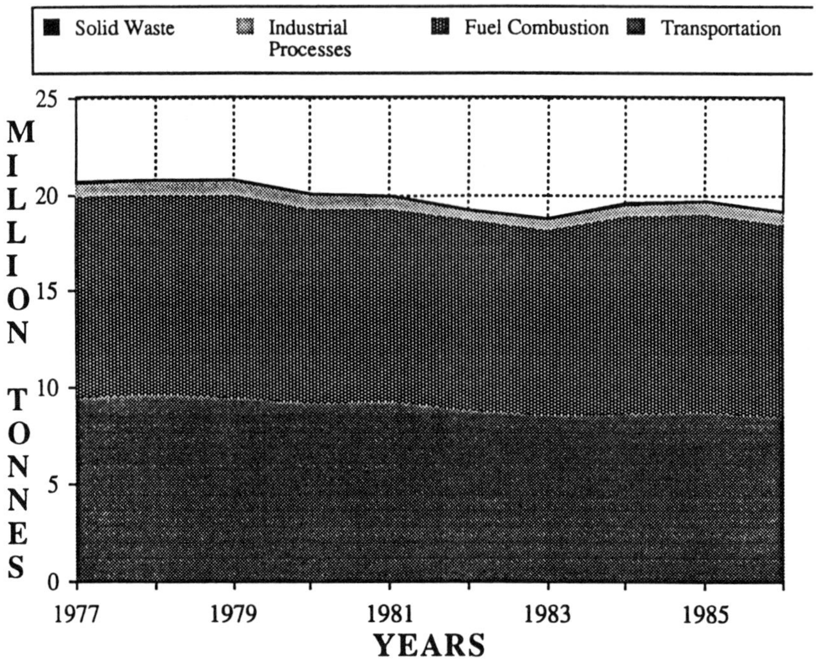

FIG 11. Emissions of nitrogen oxides in the United States from 1977 to 1986 in units of million metric tons per year (from EPA [11]).

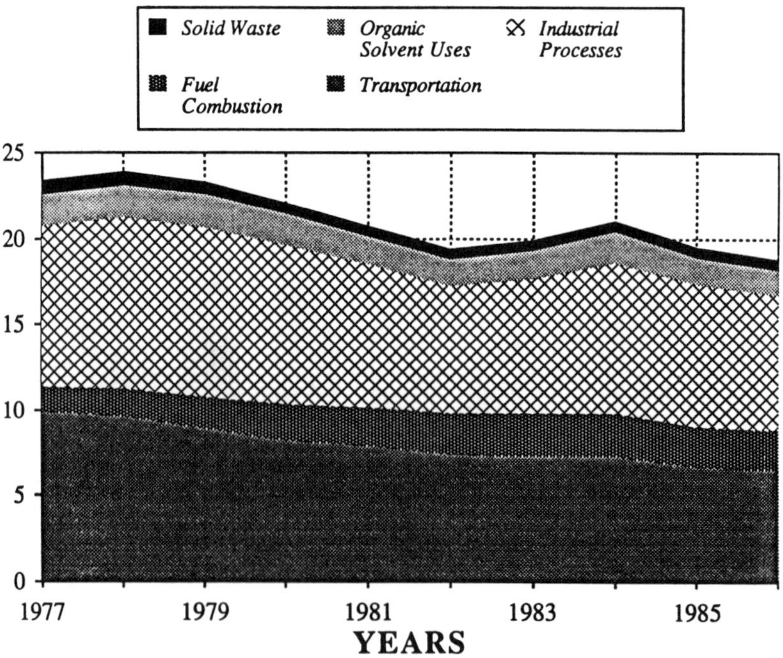

FIG 12. Emissions of volatile organic compounds in the United States from 1977 to 1986 in units of million metric tons per year (from EPA [11]).

Trends in Ambient Ozone

The number of days exceeding the NAAQS ozone standard of 0.12 ppm vary from year to year in a manner which appears to be unrelated to changes in the emission rate of ozone precursors. The temporal trends in VOC and NO_x emissions, Figures 11 and 12 respectively, as reported by EPA [11] have been relatively steady and/or decreasing over the past several years.

Over the same period there has been no consistent trends in ozone levels in the State of Michigan. Figure 13 shows the number of days with ozone levels exceeding 0.12 ppm in Muskegon, Michigan. The variations from year to year appear to be unrelated to the emission trends and much more closely aligned to the changes in mean maximum summertime temperatures. This may be because emission trends within and upwind of Michigan have not followed the national trends reported in Figures 10 and 11, or it may be because of variation in meteorological conditions to which ozone levels are sensitive.

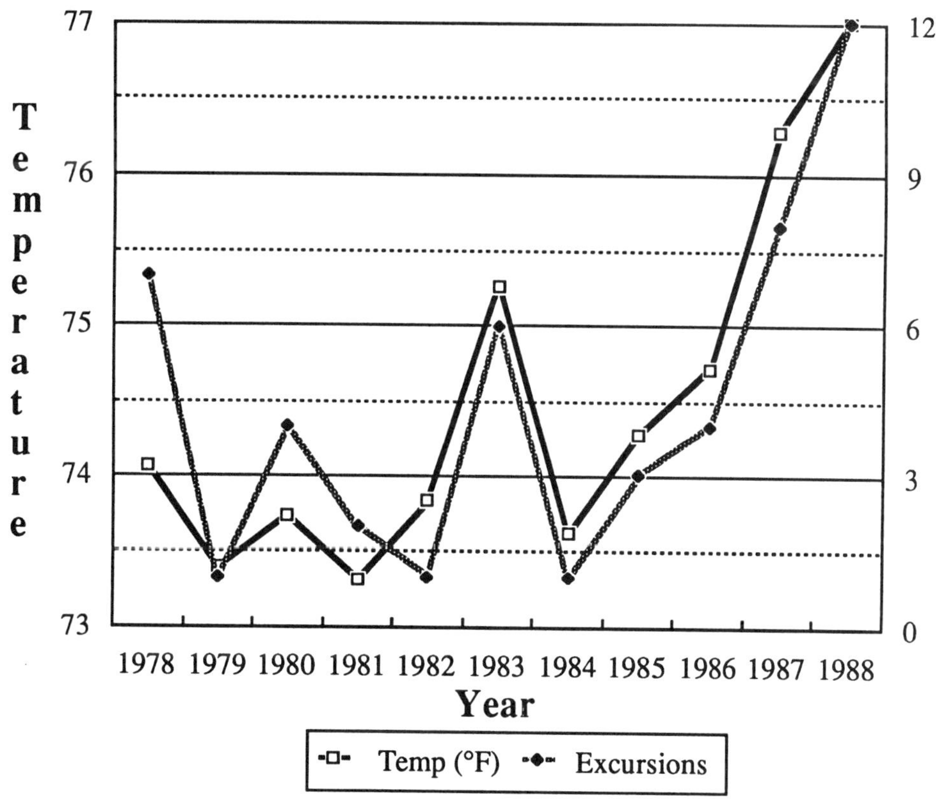

FIG 13. Temporal trends in number of days exceeding the NAAQS for ozone of 0.12 ppm as a function of mean maximum summertime (May-August) temperature.

Summary and Conclusions

Given:

- that most counties east of the Mississippi River are presently not in compliance with the NAAQS for ozone;
- the ozone exhibits a pronounced positive relationship with temperature;
- that the possibility exists for temperatures to rise with global tropospheric warming and this may lead to reduced wind speeds during more frequent heat waves;

it would be reasonable to conclude that it will be difficult to obtain the standard in the near future. Figure 14 shows the same data as a scatter diagram and illustrates that ozone exceedences increase with increasing temperature. Extrapolation to higher temperatures, as might occur with a global warming, does not suggest any reduction in ozone exceedences.

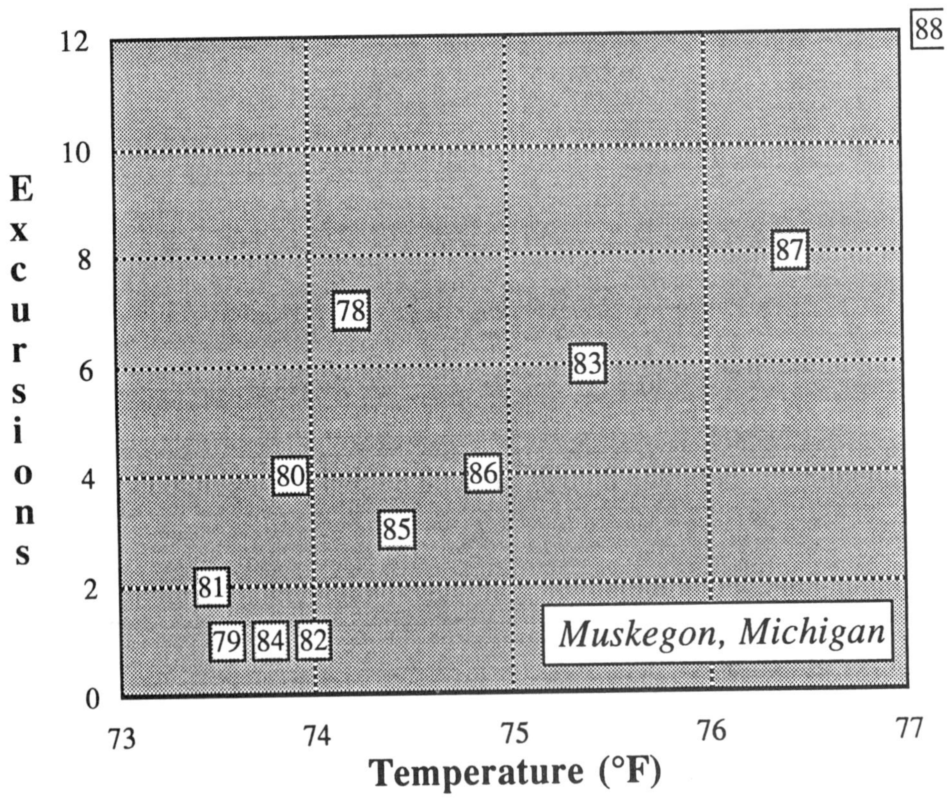

FIG 14. Scatter plot of the number of days exceeding the NAAQS for ozone of 0.12 ppm as a function of mean maximum summertime (May-August) temperature. The years are indicated in the boxes.

This information also points out that if, instead of a warming pattern, we enter an extended period of cooling due to forces not considered in this paper, it may be possible to meet the NAAQS for ozone. However, if the standard were to be met during a cooler period it would be prudent to wait until a representative set of years which includes a range of temperature conditions has passed before declaring success.

Finally, these results also suggest that, since high ozone days are most likely associated with above average temperature days, our tests of human response to ozone exposure ought to consider the combined pollutant and heat stress. It is imperative that the NAAQS for pollutants be tested under those environmental conditions under which high concentrations are most likely to occur.

References

1. National Research Council, *Carbon Dioxide and Climate: A Second Assessment* (National Academy Press, Washington, D.C. 1982).

2. National Research Council, *Changing Climate* (National Academy Press, Washington, D.C. 1982).

3. J. Hansen, I. Fung, A. Lacis, D. Rind, S. Lebedeff, R. Ruedy, and G. Russel, *J. Geophys. Res.*, (in press) (1989).

4. J.E. Penner, P.S. Connell, D.J. Wuebbles, and C.C. Covey, *Climate Change and its Interactions with Air Chemistry: Perspectives and Research Needs*, EPA Report under Interagency Agreement No. DW89932676-01-1, Lawrence Livermore National Laboratory, CA (1988).

5. J.H. Seinfeld, *Atmospheric Chemistry and Physics of Air Pollution*, (Wiley-Interscience, New York 1986).

6. B.H. Lamb, H. Westbert, G. Allwine, and T. Quarley, *J. Geophys. Res.* **90**, 2380, (1985).

7. S. Sillman, J.A. Logan, and S.C. Wofsy, *J. Geophys. Res.*, (in press) (1989).

8. S. Sillman, J.A. Logan, and S.C. Wofsy, *J. Geophys. Res.*, (submitted) (1989).

9. S.L. Grotch, *Regional Intercomparisons of General Circulation Model Predictions and Historical Climate Data*, U.S. Department of Energy, Report DOE/NBB-0084, Washington, D.C. (1988).

10. P.J. Samson, *Aggregation of Pollutant Deposition Episodes into Seasonal and Annual Estimates*, EPA Interim Report under Cooperative Agreement CR-814854-01, Research Triangle Park, North Carolina (1989).

11. U.S. Environmental Protection Agency, *National Air Quality and Emissions Trends Report*, 1986, EPA Report EPA-450/4-88-001, Research Triangle Park, North Carolina (1988).

CLIMATE CHANGE, SMOG AND ACID RAIN: LINKAGES BETWEEN POLLUTANTS, EFFECTS AND CONTROLS*

Gordon J. MacDonald

The MITRE Corporation
7525 Colshire Drive
McLean, VA 22102

Introduction

Human activity is changing the atmosphere both at an increasing rate and in increasingly complex ways. The popular terms "greenhouse," "ozone hole," "smog" and "acid rain" capture the flavor of effects brought about by man-made alterations in the chemical and radiative structures of the atmosphere. Greenhouse warming and stratospheric ozone destruction are global effects, while acid rain and urban smog tend to be regional. Even so, the atmosphere comprises a single system; reactions proceed continually over all its parts, but on different time scales. Consequently, problems related to the greenhouse, acid rain, smog and the ozone hole are strongly linked through the chemical unity of the atmosphere. For example, greenhouse warming and increased surface ultraviolet radiation generated by the destruction of stratospheric ozone promote smog formation, which, in turn, enhances and accelerates the creation and deposition of acid precipitation.

In considering the impacts of an altered atmosphere on ecosystems, the effects of global stresses cannot be separated from the stresses caused by local and regional changes. Often, stresses act synergistically, strengthening and hastening the effects of each other. Increased warmth not only intensifies urban smog, but heat and smog together have a greater health effect on a city's inhabitants than does either alone. A forest stressed by drought is more susceptible to attack by insects and more vulnerable to acid deposition and exposure to oxidants. The combined effect of the various insults may be significantly greater than the sum of the individual stresses. Climatic, radiative, and pollution stresses on individual species and on limited ecosystems have been examined, but studies of interactive effects are largely limited to the combined effects of various air pollutants.

*This paper, which was prepared for the conference, was not presented at that time due to scheduling conflicts.

Copyright 1989 by Elsevier Science Publishing Company, Inc.
Global Climate Change Linkages
James C. White, Editor

For historical reasons, the legal basis in the United States for dealing with air quality problems—the Clean Air Act—has been focused on single pollutants and their effects. The regulation of ozone is an exception owing to the recognition that reactive hydrocarbons and oxides of nitrogen are essential to ozone formation; the Clean Air Act does provide for restrictions on these precursors to ozone. In recent years, a variety of measures that have the potential for worsening other problems have been proposed to deal with sulfur oxide. Strict emission regulations on sulfur oxide emissions from coal-burning power plants have been proposed (for example, in the Clean Air Standards Attainment Act of 1987, S.1894). The scrubbers required to remove sulfur dioxide from flue gas reduce the overall efficiency of conversion of coal to electricity. As a result, more greenhouse carbon dioxide is emitted into the atmosphere per unit of useful energy generated. Moreover, because sulfur dioxide acts as a sink for ozone and other atmospheric oxidants, reduction of SO_2 concentration may escalate oxidant damage to forests.

Other proposed measures aimed at reducing urban smog have unintended consequences, as well as enhancing greenhouse warming by increasing lower-atmospheric ozone. Methanol has been advanced as an alternative automobile fuel to ease air pollution, principally carbon monoxide, in congested urban centers. Burning methanol derived from natural gas places slightly more carbon dioxide in the atmosphere per mile of transport delivered than does gasoline, and 50% more than compressed natural gas, another alternative automotive fuel. The emphasis in the several alternative fuel proposals has been to solve the immediate, regional problem and not to consider other issues raised by the purported solution.

The analysis of interactions among chemical species during smog formation extends back to pioneering work done by Haagen-Smit in the late 1940s (Haagen-Smit, Bradley, and Fox, 1952; Haagen-Smit, 1952). More recently, analysis of the interaction of individual pollutant species has been the focus of atmospheric chemistry (Finlayson-Pitts and Pitts, 1986; Seinfeld, 1986). Yet the potential effects on ecosystems stemming from the interplay of local polluters with global changes in climate and ultraviolet radiation have received scant attention. Even less analytic effort has gone into examining the unintended consequences of various control schemes legislated to remedy immediate pollution problems. Analyzing the links between atmospheric chemistry, the effects of atmospheric constituents, and control strategies is of critical importance.

The purpose of this paper is to illustrate the extensive interactions among atmospheric constituents, the importance of synergetic effects, and the need to take into account the interactive nature of the atmosphere and biosphere when framing legislation. I first describe the atmosphere as a dilute combustion chamber in which man is altering both the fuels and oxidizers. Oxidants, with their complex chemistry, play a key role in acid precipitation and greenhouse warming, as well as damaging ecosystems directly. Secondly, I will consider the various pathways by which environmental changes will perturb one important ecosystem, the temperate forest. I conclude with preliminary thoughts on possible integrated approaches to minimizing future adverse alterations to the atmosphere. My emphasis on the atmosphere is a first step toward forming a comprehensive view of the environment that would include water and land.

The Changing Atmosphere

The Atmospheric Combustion Chamber

Not many of us recognize it, but we live in a dilute, low temperature combustion chamber—the atmosphere, as symbolically illustrated in Figure 1. The fuels are compounds of carbon, nitrogen, sulfur and hydrogen, such as nitric oxide (NO), carbon monoxide (CO), methane (CH_4) and sulfur dioxide (SO_2), and are emitted by a variety of natural and anthropogenic sources. Photochemical reactions are driven by solar ultraviolet radiation that filters through the upper atmospheric ozone layer. The intensity of the ultraviolet radiation determines in part the rate at which oxidants form. As upper ozone is removed, lower-atmosphere oxidant formation is stepped up, producing a more oxidizing atmosphere. The major oxidants are ozone (O_3), hydroxyl radical ($HO\cdot$), hydroperoxyl radical ($HO_2\cdot$), hydrogen peroxide (H_2O_2), nitrate radical ($NO_3\cdot$), organic peroxyl radicals ($RO_2\cdot$), and peroxyacetyl nitrate (PAN). In both gas and liquid phases (cloud and fog droplets), oxidants react with fuels (see Figure 1) to form long-lived species such as carbon dioxide (CO_2), or short-lived, acidic species such as nitric (HNO_3) and sulfuric acids (H_2SO_4).

Oxidants differ in their reactivity. In the gas phase, hydroxyl radical and ozone are the first and second most reactive oxidants. Hydrogen peroxide and PAN do not directly oxidize in the gas phase, but serve as important reservoirs for hydroxyl and organic peroxyl radicals. Hydrogen peroxide, a strong oxidizer in the aqueous phase, becomes more potent than ozone as the pH of the liquid drops below five.

As will be discussed, oxidants are of particular importance to a consideration of effects on the biosphere; they cause direct damage, as well as provide a pathway for the formation of damaging acids. The atmospheric chemistry of oxidants is complex (see, for example, Seinfeld, 1986). The Environmental Protection Agency (EPA), however, has standards only for ozone. EPA argues that ozone is a satisfactory regulatory surrogate for all oxidants. Regulation of ozone is intended to regulate other photochemical oxidants indirectly, even though the chemical routes and reactivity of these other oxidants differ from ozone.

Origin of Ozone

Only one reaction forms ozone in the lower atmosphere—the addition of an oxygen atom to an oxygen molecule in the presence of a third molecule:

$$O_2 + O \rightarrow O_3 + M$$

The oxygen atoms are derived primarily from the photolysis of nitrogen dioxide (NO_2) according to

$$NO_2 + h\nu(\lambda < 398\, nm) \rightarrow O + NO$$

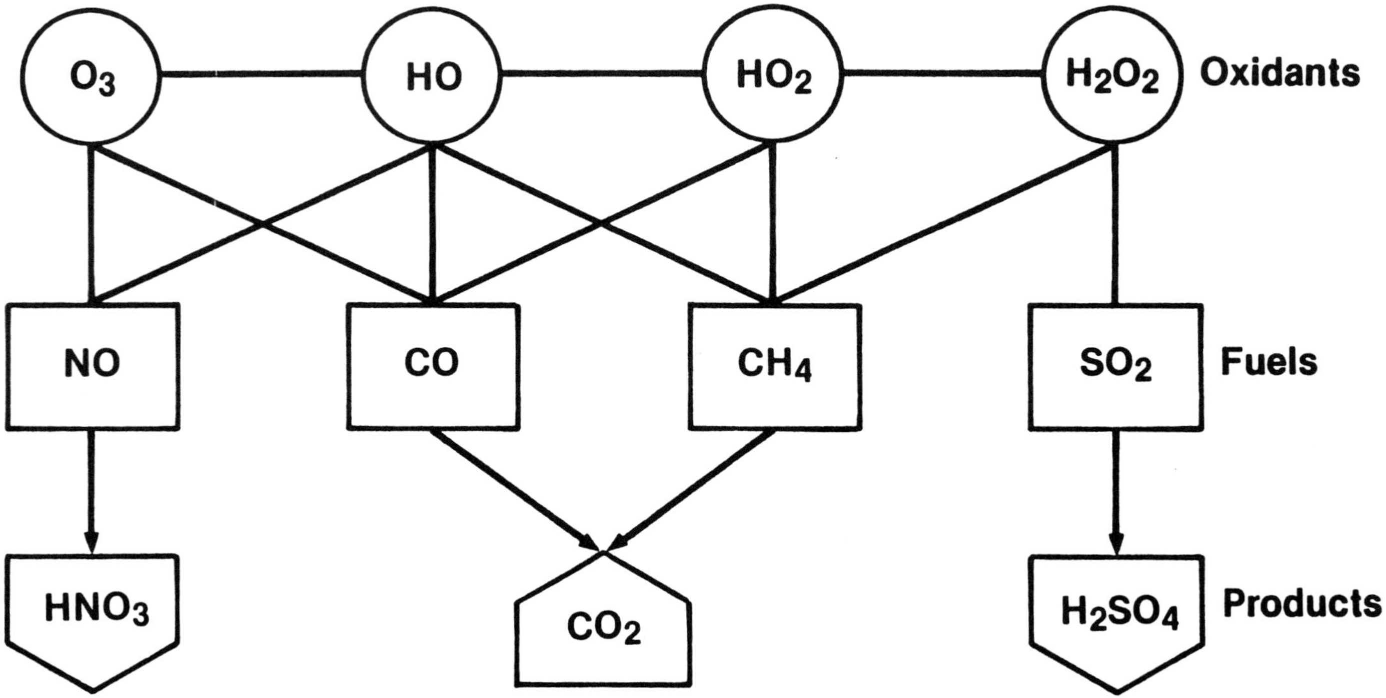

Figure 1
Schematic Illustration of the Atmosphere as a Low Temperature, Dilute Combustion Chamber

The fuels are man-made and the oxidants are produced by solar radiation, while the products give rise to problems in climate and acid deposition.

where *hv* denotes a photon of ultraviolet radiation. One further reaction completes the "nitrogen cycle" for ozone formation. This reaction is the molecule-molecule exchange between ozone and nitric oxide, which reforms nitrogen dioxide:

$$O_3 + NO \rightarrow NO_2 + O_2$$

The transient concentration of ozone is thus fixed by the ratio of nitrogen dioxide to nitric oxide according to

$$[O_3] \cong 0.021\, ppm \frac{[NO_2]}{[NO]}$$

where [ ] denotes concentration. During the day, with its flux of solar ultraviolet radiation, there is a continuous cycle of formation and destruction of ozone, accompanied by the conversion of nitrogen dioxide to nitric oxide. The equilibrium concentration of ozone in these circumstances is fixed by the *ratio* of NO_2 to NO in the background atmosphere. Indeed, Liu et al. (1987) find that both observation and model calculations show daily ozone production per unit of nitrogen oxide (NO_X) to be greater in rural areas with lower NO_X concentrations. Thus, it is not the total concentration of nitrogen oxides that determines ozone concentration, but the ratio of the two nitrogen oxide species. This ratio is fixed by the complicated photochemistry of hydrocarbons and carbon monoxide – a fact that has important implications for regulatory strategies.

In addition to natural sources of nitrogen oxides, combustion produces both NO_2 and NO, generally in the ratio of 1:20 or less depending on the temperature of combustion. At these ratios, the background concentration of ozone is small. Indeed, if the chemistry involved only sunlight, nitrogen oxides and background chemistry, there would be no difficulty in meeting the current EPA standard (a one-hour average concentration of O_3 less than 0.12 ppm) for the protection of human health. An additional path for the conversion of NO to NO_2 that does not destroy ozone is required for the ozone concentration to build to those levels harmful to health. This alternate path is provided for by the photochemical oxidation of reactive hydrocarbons and by carbon monoxide.

The key reaction in hydrocarbon oxidation that converts NO to NO_2 is the fast radical transfer reaction with NO of the form:

$$RO_2 \cdot + NO \rightarrow RO \cdot + NO_2$$

where $RO_2\cdot$ is an organic peroxyl radical, and R can be hydrogen or virtually any organic moiety or fragment. Carbon monoxide participates in the conversion of NO to NO_2 through the chain reaction

$$CO + HO\cdot \rightarrow CO_2 + H\cdot$$

$$H\cdot + O_2 + M \rightarrow HO_2\cdot + M$$

$$HO_2\cdot + NO \rightarrow HO\cdot + NO_2$$

$$NO_2 + h\nu \rightarrow NO + 0$$

$$O + O_2 + M \rightarrow O_3 + M$$

The net reaction for this cycle is

$$CO + 2O_2 \rightarrow CO_2 + O_3$$

In this chain, the hydroperoxyl radical $HO_2\cdot$ plays the role of $RO_2\cdot$. Control of ozone concentration thus requires limits on emissions of nitrogen oxides, reactive hydrocarbons, and carbon monoxide.

Methane, a strong greenhouse absorber, also participates in ozone formation. The oxidation of CH_4 in the presence of adequate NO_x (urban setting) produces ozone and can also produce $HO\cdot$. Crutzen (1987) calculates that in smog-forming conditions, oxidation of methane to CO_2 and H_2O produces 3.7 O_3 molecules and 0.5 $HO\cdot$ radicals per each methane molecule destroyed. However, in large fractions of the troposphere, NO_x mole fractions are probably 10 ppt or less. In these circumstances, methane oscillation consumes ozone, $HO\cdot$ and $HO_2\cdot$ in producing CO_2, H_2O, and H_2. Crutzen (1987) estimates that under these low NO_x concentrations, the oxidation of each CH_4 molecule consumes 1.7 O_3 molecules and 3.5 HO_x molecules.

This, then, is the paradoxical situation within the atmosphere: nitrogen oxides are essential to ozone formation, even though their total concentration does not determine ozone levels. Of course, if NO_x could be eliminated, ozone levels would be extremely low. Unfortunately, natural sources of NO_x do not permit such a clean solution. Other nitrogen compounds involved in the complex chemistry, particularly PAN and nitrate radical, serve to muddy the above generalization, even in urban settings. But these are details. The most important message gleaned from a consideration of ozone formation is that, in order to reduce ozone levels in the urban setting, one has to reduce the levels of hydrocarbons *and* carbon monoxide.

Other Strong Oxidants

The production of oxidants other than ozone involves reactions with ultraviolet waves less than 315 nm in length. The photodissociation of ozone,

$$O_3 + h\nu(\lambda < 315\,nm) \rightarrow O(^1D) + O_2$$

generates excited-state oxygen atoms $O(^1D)$. A small fraction of the excited oxygen atoms, about 1%, reacts with water vapor to produce the hyperreactive hydroxyl radical:

$$O(^1D) + H_2O \rightarrow 2OH\cdot$$

The reactions that lead to hydroxyl constitute a major photochemical sink for ozone. This sink limits the lifetime of ozone in the lower atmosphere to about five days (Liu et al., 1987). Ozone is also removed by solution and oxidation by SO_2—processes that reduce ozone lifetime to two days. In winter, both the photochemical and surface-deposition sinks diminish, and ozone lifetime extends to more than a month.

The hydroxyl radical is a strong oxidizer essential to the formation of other oxidants. Hydroperoxyl radical forms through the interaction of hydroxyl and carbon monoxide according to

$$HO\cdot + CO \rightarrow CO_2 + H\cdot$$

$$H\cdot + O_2 \rightarrow HO_2\cdot$$

Hydroperoxyl radicals combine to form hydrogen peroxide:

$$HO_2\cdot + HO_2\cdot \rightarrow H_2O_2 + O_2$$

The extreme reactivity of hydroxyl limits its concentration in the atmosphere. Because of its sparsity, the direct measurement of hydroxyl concentration in the atmosphere has proved elusive. Perner et al. (1987) have used ultraviolet absorption spectroscopy to obtain values of 0.7 to 3.2 x 10^6 molecules cm^{-3}.

Hydrogen peroxide is another important oxidizing agent, partly because of its high solubility in liquid water. High hydrogen peroxide concentrations are favored by high concentrations of carbon monoxide and volatile organic compounds, and low concentrations of nitrogen oxide. High NO_x levels lead to the formation of nitrate radicals from organic peroxyl radicals, and H_2O_2 production is suppressed. Present observations (Heikes et al., 1987) show H_2O_2 varying between 0.2 to 4.1 ppbv in the lower 3.5 km of the atmosphere, with the lowest values near the ground and the highest values at altitudes just above cloud tops.

Role of Oxidants in Acid Deposition

Nitric acid and sulfuric acid are the strong acids implicated in damage associated with acid deposition. The precursors of these acids, SO_2 and NO_x, are emitted into the atmosphere in roughly equal tonnages. The impacts of sulfuric and nitric acid differ, however. A unit mass of SO_2 has the potential of producing 44% more acidity (H_2SO_4) than a unit mass of nitrogen dioxide, which forms HNO_3 with a single hydrogen ion. In describing the damaging effects of acids, it is the concentration of hydrogen ion or pH (the logarithmic measure of hydrogen ion concentration) that is significant.

In addition to sulfur dioxide and nitrogen oxides, oxidants are an essential precursor to nitric and sulfuric acids. Current proposals to deal with the acid rain problem involve limitations on emissions of NO_x and SO_2, but even when dissolved in water, these substances are relatively harmless unless oxidized. Sulfur dioxide must remain in a cloud for several days before ambient oxygen converts it to sulfuric acid. In the short time (minutes to hours) that precursors normally remain within a cloud, strong oxidants are required to form acids. This conclusion is supported by evidence from studies of atmospheric chemistry, by the observed variation in intensity of acid precipitation from winter to summer, and by historical variation of acid deposition.

In the gaseous phase, NO_x and SO_2 behave very differently. Ozone converts NO to NO_2 rapidly. The NO_2 is then attacked by hydroxyl, which converts it to HNO_3, with a mean reaction time of several hours. The equilibrium density of NO_2 in water, relative to air, is almost 10^7 more than the relative volume of air to water droplets in a typical rain cloud. Therefore, after insertion into an air parcel, NO_x can be transformed to nitric acid within the existing water droplets, or within droplets that subsequently form, in less than a day.

Sulfur dioxide has a tendency to react with oxygen in air:

$$2SO_2 + O_2 \rightarrow 2SO_3$$

When formed, SO_3 reacts so rapidly with water to form sulfuric acid,

$$SO_3 + H_2O \rightarrow H_2SO_4$$

that, in a moist atmosphere, any process that forms SO_3 will inevitably form sulfuric acid. However, the key reaction of SO_2 with O_2 is so slow without the presence of a catalyst, that this reaction can be totally disregarded as a source of sulfuric acid.

By far, the most important gas-phase oxidation reaction for SO_2 involves hydroxyl:

$$SO_2 + HO\cdot + M \rightarrow HOSO_2\cdot + M$$

The $HOSO_2\cdot$ radical ultimately converts to H_2SO_4, though the detailed mechanism remains uncertain. In the process, HO· is regenerated, so that in a photooxidizing mixture of HNO_2, NO, NO_2, and CO, the concentration of HO· remains constant, even with large additions of SO_2 (Meagher, Olszyna and Luria, 1984). This observation is important. If gas phase oxidation by HO· dominates, a reduction of SO_2 would result in a proportionate reduction in sulfuric acid deposition. The reaction of SO_2 with HO·, the rate-limiting process, is very slow. On a cloudless summer day with typical concentrations of the relevant species, only some 16% of the SO_2 present in the atmosphere would be converted to H_2SO_4 over a twenty-four hour period. In wintertime, the fraction converted would be even less, because of the lower HO· levels resulting from the seasonal reductions in ultraviolet radiation.

In the liquid phase, the high solubility of SO_2 in water causes the formation of H_2SO_4 from SO_2 to proceed far more rapidly than in the gaseous phase, provided a strong oxidant is available. Dissolved ozone is the dominant oxidant if the water droplets have a pH greater than 5 (Heikes et al., 1987). With higher acidities (pH less than 5), hydrogen peroxide becomes the leading oxidant of dissolved SO_2. In the eastern United States, the pH of atmospheric water (rain, snow, cloud droplets, fog, and haze) is generally below 4.3. In such circumstances, the amount of SO_2 converted to H_2SO_4 will be regulated by the amount of hydrogen peroxide that is available. As has been noted, the production of hydrogen peroxide depends on the availability of hydrogen atoms, H·. The concentration of H· is determined by HO·; the concentration of HO· depends both on ultraviolet radiation and on the availability of carbon monoxide. CO thus plays a key role in forming the two strong oxidants, O_3 and H_2O_2, required to convert SO_2 to sulfuric acid.

The differing oxidation paths of NO_x and SO_2 are reflected in the differing seasonal variations of deposition for nitric and sulfuric acid. Although emission rates of both gases remain approximately constant over the year, the seasonal variation in sulfate deposition is much greater than that of nitrate in the eastern United States, as is illustrated in Figure 2 (after Henderson and Weingartner, 1980). In any given parcel of air, the conversion of NO_x to HNO_3 takes place far more rapidly than does the oxidation of SO_2 to H_2SO_4. It would appear that NO_x uses up gas-phase oxidant first, and only the remaining oxidant is available to convert SO_2 to sulfuric acid, primarily in the liquid phase. In wintertime, when ultraviolet and oxidant levels are low, levels of sulfuric acid deposition are also lower. These observations, together with chemical considerations, lead to the conclusion that sulfuric acid deposition is limited by the availability of oxidants in winter. Reduction of SO_2 emission alone during that season may not result in any lowering of acid deposition. Yet wintertime deposition stores acid in snow, to be released abruptly in the spring melt, with large adverse effects on aquatic life. Because

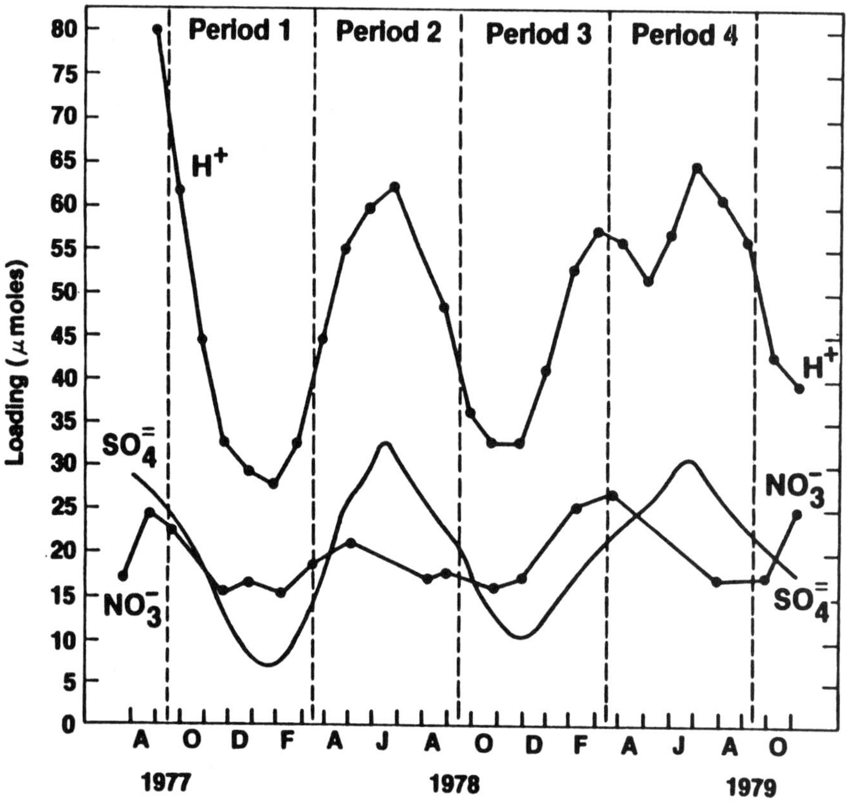

Figure 2
Ion Loading to Precipitation in Ithaca and Whiteface, New York and State College, Pennsylvania

Loading of nitrate, sulfate, and hydrogen ions from August 1977 to October 1979 (after Henderson and Weingartner, 1980).

only a fraction of the available SO_2 is deposited as acid in the summer, sulfuric acid deposition may be oxidant-limited during this season as well.

The production of nitric acid is not seasonally limited. Hydroxyl appears to be in sufficient abundance, even during winter, to convert NO_x to nitric acid in the gas phase; reduction of nitric acid deposition can only be brought about by reduction in NO_x emissions. Because gas phase oxidation of SO_2 by hydroxyl proceeds far more slowly than oxidation of NO_x, oxides of nitrogen—with their lower potential to produce damaging hydrogen ions—are precipitated out first, leaving unoxidized SO_2 in the traveling air mass. Liquid-phase oxidation of SO_2 with low pH (less than 5) requires hydrogen peroxide. The formation of hydrogen peroxide depends directly on the presence of carbon monoxide, which combines with hydroxyl to provide the essential hydrogen for the formation of hydroperoxyl and, subsequently, hydrogen peroxide. Carbon monoxide is generated by both anthropogenic sources and natural processes, such as the atmospheric oxidation of methane by hydroxyl.

Greenhouse Warming

Human alteration of atmospheric composition disrupts the radiative balance of the atmosphere. These chemical and radiative changes can lead to long-term shifts in climate, primarily by deepening absorption in the 8 to 12 μm infrared region—the "window" of the atmosphere, which today is partially transparent to escaping radiation. Carbon dioxide has long been known as an infrared absorber that reduces the window's transparency, and the term greenhouse is usually associated with atmospheric warming due to increased CO_2 levels. Atmospheric carbon dioxide levels have increased from preindustrial levels of about 280 ppmv to the present concentration of 350 ppmv—a 25% rise (From and Keeling, 1986). But carbon dioxide is not the only radiatively active gas of anthropogenic origin. Concentrations of other heat-absorbing gases are also on the rise; methane and nitrous oxides have been increasing recently at the rates of about 1% and 0.3% per year, respectively (Ramanathan et al., 1985). Concentrations of low-level ozone in Europe have risen from 10 to 20 ppbv, during the early part of this century, to the present annual average values of 25 to 35 ppbv (Crutzen, 1988). Emissions of chlorofluorocarbons (CFC-11, 12), which have no natural sources, have increased by an order of magnitude over the last 30 years. CFC emissions began leveling in 1974, but more recently the increase has resumed (Gamlen et al., 1986). Emissions of other radiatively active trace gases are also growing.

Changes in global average temperature due to perturbation of the radiative properties of the atmosphere have been analyzed using a variety of models ranging in complexity from simple energy budget models to complex global consolidation models (GCMs). The models agree that, of the anticipated warming, about half will be generated by carbon dioxide, with the other trace gases accounting for the rest (Ramanathan, 1988). For example, the infrared heating effect produced by an increase in methane from 0.7 ppm (preindustrial revolution) to 1.7 ppm (1988 concentration) is about half as large as the effect of a simultaneous increase in the CO_2 concentration from 280 ppm to 350 ppm.

Trace gases interact both radiatively and chemically to alter climate. Through their chemical reactivity, CFCs are responsible for the ozone hole in Antarctica (Farman, Gardiner, and Shanklin, 1985; Stolarski et al., 1986), and may lead to a global reduction in stratospheric ozone. CFCs are also radiatively active as very strong absorbers of infrared; their net effect on warming is due both to the catalytic reduction of ozone and absorption of outgoing infrared radiation. Methane, too, acts as an infrared absorber and, in the stratosphere, it interferes with the ozone-depleting effects of the chlorine released by photodecomposition of CFCs. The interactive role of methane is further extended by its potential effects on water formation in the stratosphere, where photodissociation of CH_4 is a major source of hydrogen and water vapor. Ice crystals derived from water vapor provide surfaces for the heterogeneous reaction of ozone with chlorine, thus enhancing the ozone-reducing potential of the chlorine released by photodissociation of CFCs. The oxidation of methane throughout the atmosphere contributes about 6% of the annual atmospheric increase in CO_2.

Tropospheric ozone accounts for only 10% of the typical column of ozone, but pressure broadening significantly increases its infrared absorption. As has been noted, ozone also initiates reactions that generate the highly chemically active oxidants, including hydroxyl. The lifetime of radiatively active methane in the atmosphere is determined by hydroxyl abundance. Carbon monoxide and methane are both sinks for hydroxyl, CO being the more reactive of the two. Greater concentrations of carbon monoxide in the atmosphere lead to lesser concentrations of HO·, which in turn lead to higher concentrations of methane.

Carbon monoxide and ozone are linked in the non-urban atmosphere that is low in hydrocarbons through interaction with HO·, as described above, and NO_2. Hydroxyl is destroyed by CO, increasing the ambient concentration of NO_2. The resulting increased proportion of NO_2 to NO leads to a higher steady state concentration of ozone. Carbon monoxide, a non-radiatively active species, enhances the greenhouse warming by raising the levels of methane and ozone.

The correlative nature of ozone and carbon monoxide concentrations in the troposphere is well established. In northern regions of industrial activity, concentrations of O_3 and CO at remote sites are higher than at similarly isolated sites in the southern hemisphere (Fishman and Seiler, 1983). If the concentrations of CO and O_3 are measured as a function of altitude, O_3 concentrations tend to be higher at levels of high CO concentration. The only locations where the correlation is not positive are those where there has been a down-mixing of stratospheric ozone into the troposphere. These observations are consistent with tropospheric chemical models in which ozone is produced not only in urban regions, but also in rural areas, as a result of elevated levels of CO and natural hydrocarbons.

The spatial association of methane with carbon monoxide is not as well determined as that of ozone. Measured methane concentrations have increased about 1% a year over at least the past three decades, even though there has been no parallel increase in established sources of methane. Because the heightened CH_4 mimics the increases in anthropogenic CO, it is reasonable to assume that at least some fraction is added by the destruction of methane's principal sink, hydroxyl radical, by carbon monoxide.

Combined Effects of Stratospheric Ozone Depletion and Greenhouse Warming on Urban Smog and Acid Rain

Ozone has a strong ultraviolet absorption band that extends up to 340 nm; its presence in the upper atmosphere limits the transmission of ultraviolet radiation to the surface. A removal of stratospheric ozone will allow more ultraviolet to penetrate to lower levels, particularly in the 295 to 315 nm region. A greater influx of radiation will cause significant increases in the photolysis rates of those species that absorb in this region, specifically ozone, and selected hydrocarbons such as formaldehyde and aldehydes of higher molecular weight. In general, an increase in ultraviolet will speed up the formation of photochemical smog.

Photolysis reaction rates have a functional dependence on temperature of the form exp(-E/kT), where E is the activation energy and k is Boltzman's constant. For a small change in temperature (a few degrees), the increase in reaction rate will be small as well (a few tenths of a percent to a few percent). For example, the reaction rate for

$$NO + O_3 \rightarrow NO_2 + O_2$$

increases by 8% for a five-degree increase in temperature. Warming, then, will also generally speed up formation of photochemical smog by a small fraction.

The stepped-up photolysis of ozone under heightened ultraviolet radiation increases the rate of production of excited oxygen atoms, which react rapidly with water vapor to form hydroxyl. Higher hydroxyl levels promote the formation of both hydroperoxyl and hydrogen peroxide, the latter of which is the most likely oxidizer of SO_2 to sulfuric acid. However, quantitative calculation of the combined effects of increased temperature and ultraviolet flux is required to estimate the magnitude of said effects on urban and regional pollution.

Effects of Changing Atmospheric Composition on Temperate Forests

Introduction

The forests of the temperate zone in the Northern Hemisphere comprise complex systems with many possible responses to changing atmospheric composition, both in terms of direct effects from airborne pollutants and ultraviolet radiation, and indirect effects associated with changing climate. Forest ecosystems are driven directly and indirectly by solar energy; the quantities of energy involved are huge. In New England alone, natural ecosystems consume about 87 quads of energy per year – roughly 30 times the amount used by humans in that same area (Bormann, 1982).

The last three decades have seen extensive studies on the decline of forests in western Europe and eastern North America. Because the forest systems are complex and vulnerable to a wide variety of natural and anthropogenic changes, determining the cause of change is extremely difficult. Consequently, the establishment of a clear, cause-and-effect relationship between observed ecologic changes and changes in atmospheric chemistry may not be possible.

The atmosphere interacts with a forest system in numerous ways. In the gaseous phase, interactions occur directly with the foliar structure and with soils and water. Precipitation can interact directly with a tree and its roots, as well as chemically alter the soils and microorganisms within the soils. The recognition that industrial emissions could cause forest damage came in the 19th century as plant "deserts" developed around industrial centers in England, such as Manchester and Liverpool. Similar damage was noted in the valleys of the Erzgebirge in Germany at the end of the 19th century. Evidence indicated that conifers were sensitive to long-term exposure to low pollutant concentrations, and deciduous trees appeared more sensitive to high, intense doses. The pollution was attributed to smoke, though there was an early recognition that acidification of soils through the deposition of sulfuric acid might play a destructive role (Wieler, 1911).

Modern analysis of the effects of air pollution on forests began in the Los Angeles basin during the late 1940s. Ponderosa and Jeffrey pines in the San Bernardino mountains were found to be highly susceptible to smog attacks, while white fir and black oak suffered less damage. The damaged trees were more susceptible than healthy trees to insect attack and root rot. Controlled experiments with young pines implicated ozone as the principal damaging agent (Miller et al., 1969). However, the recent discovery that highly acid fogs are found in the Los Angeles basin (Munger et al., 1983) raises the possibility that the damage observed in this classic case may be due to a combination of acid and oxidant stress. Further, the potential importance of hydrogen peroxide as a damaging agent had not yet been recognized at the time the San Bernardino forest was extensively investigated. This case illustrates the complexities involved with assigning specific responsibility for adverse effects to a single pollutant. In a review of forest damage, McLaughlin (1985) concludes, "There are still many postulated pathways and mechanisms for observed responses, including both direct and indirect influences of O_3, wet and dry deposited strong acids, and heavy metals, as well as climatic change... However, there is still no clear evidence of a single causal agent."

Specific Cause-and-Effect Relationships

A number of effects produced by alterations in atmospheric composition and radiation levels have been identified, mostly in connection with the impacts of pollutants on forests in eastern North America and western Europe. These mechanisms include:

1. Gas-phase interaction of oxidants and other components with foliage;

2. Effects of acid precipitation, and cloud and fog water droplets on foliage;
3. Mobilization of metals in soils by acid deposition, with resulting impacts on root uptake and nutrient impoverishment;
4. Deposition of metals in soils by precipitation;
5. Nitrogen and sulfur fertilization;
6. Carbon dioxide fertilization;
7. Alteration of climatic conditions, particularly temperature and precipitation.

The effects of these man-made changes are compounded by a variety of natural influences that may be intensified by human activity: disease, insect infestations, forest fires, natural nutrient deficiencies or excesses, natural weather extremes such as droughts, high or low temperature extremes, high winds, and natural competitive stresses.

The direct effects of gaseous ozone have been extensively studied. Ozone reduces net photosynthesis in leaves or needles prior to their exhibition of observable damage (Reich, 1987). Ozone also disrupts stomata-controlled respiration and enzyme function to greatly varying degrees among different species. Heavy sulfur dioxide emission near smelters has led to extensive studies of SO_2 damage to plants; the major effects are related to enzyme damage and foliar cell destruction. Experimental evidence on the combined effects of O_3 and SO_2 is limited, but the indications are that the effects are synergistic (McLaughlin, 1985; Chappelka, Chevone, and Bank, 1985).

Although there is clear, controlled experimental data showing how ozone can damage trees, field observation suggests that ozone alone or in combination with SO_2 is not solely responsible for observed damage. For example, ozone levels in the Netherlands are higher than those in severely affected sections of the Black Forest. Ozone concentrations do not exhibit sharp spatial gradients, whereas local areas can show significant differences in the number of afflicted trees.

Hydrogen peroxide is known for its phototoxicity and is found in high concentrations in fog and cloud water (Heikes et al., 1987). Masuch et al. (1986) found that six weeks of daily exposure (three hours) to hydrogen peroxide-bearing acid fog (1-5 ppm) produced serious effects on the leaves and needles of young beech and spruce trees. The observed symptoms suggest that damage to the internal structure of needles and leaves decreased the trees' ability to transport assimilates. Because exposure to cloud droplets is dependent on tree location, the inherent patchiness of fog may explain the irregularity of tree damage. Experimental data are insufficient to judge whether or not there is a synergistic effect between acids and strong oxidants.

Deposition of heavy metals from the atmosphere may also affect vegetation. The transport and availability of heavy metals for root uptake would be increased through acidification and oxidation, particularly by highly water soluble H_2O_2. Even so, no specific link with observed forest decline has been proposed for deposited metals.

The effects of regional pollutants on temperate forests are poorly understood, but those effects associated with global atmospheric changes have received even less attention. Under controlled conditions, high ambient concentrations of carbon dioxide have enhanced CO_2 assimilation and, consequently, growth rates of individual trees over the short term (Shugart et al., 1986). It is highly uncertain that these limited experimental results can be extended; whether or not such effects would be sustained over any period of time, and whether they would improve productivity in actual forest environments is unknown. In natural environments, a rise in CO_2 concentration may aggravate natural competition as various species respond differently to the increase. If elevated CO_2 levels do result in enhanced long-term growth, increases in productivity are more likely to occur on tree farms rather than in mature forests where the capacity for carbon storage is more limited. A major problem in evaluating the impact of higher CO_2 concentrations is the impracticability of conducting controlled experimentation. The impact of the natural CO_2 increase has not been evaluated because of the impossibility of maintaining an appropriate control.

The effects of changes in climate on temperate forest productivity are also difficult to assess. Forests are sensitive to precipitation and extremes in weather, as well as to temperature. Climate models make relatively robust estimates of future temperature change, but these models give little information about possible alterations in precipitation patterns, and no information on the statistics of extreme events.

Increases in ultraviolet radiation represent yet another stress on vegetation (Caldwell, 1979). One major effect of such an increase appears to be an inhibition of photosynthesis; changes in patterns of competition and yield reductions are also observed. The effect produced by the interaction of these changes with the stresses due to climate change or pollution is not known.

Linkage of Effects

The preceding discussion emphasizes the lack of adequate data on the effects that changes in the chemical and radiative properties of the atmosphere have on forests. A number of possible linkages among the various stresses have been noted above. One potentially important linkage involves the interaction of enhanced oxidant abundances with other air pollutants, both chemically and biologically. Increases in urban-generated ozone, coupled with enhanced ultraviolet radiation, will promote the formation of hydrogen peroxide. Hydrogen peroxide and other oxidants can oxidize dry deposited SO_2 and thus accelerate the in situ formation of acid in droplets on plant foliage. H_2O_2 itself is known to be phytotoxic, but its overall effect on natural ecosystems is just beginning to be studied. Linkages between regional and global effects are similarly uncertain. For instance, increases of carbon dioxide may increase stomatal resistance in plants, making pollutant damage less likely (Pittock and Nix, 1986). However, because of the very high water solubility of H_2O_2, this argument may not apply to hydrogen peroxide.

Changes in the chemical and radiative properties of the atmosphere may substantially alter forest ecosystems (and others) over the next decade. Evidence is accumulating on the nature of regional alterations, but far less is known about the future adjustments in ecosystems that will result from climate change. Enhanced oxidant levels are likely to play a greater role in damaging both natural and managed ecosystems. As oxidants become more abundant, particularly ozone and hydrogen peroxide, they will:

1. Interfere with the natural sulfur cycle, leading to acidification of soils and water;
2. Increase mobilization and transport of potentially damaging metals in soils;
3. Act directly as phytotoxins;
4. Enhance carbon and nitrogen respiration from soils by accelerating decomposition.

Global changes will interact with regional changes. Warming may accelerate pollutant damage, while increased carbon dioxide may reduce pollutant stress.

Control Strategies

Concern about the direct human health effects of urban smog led the United States to enact its first modern environmental law in 1963 – the Clean Air Act. This law and its subsequent amendments provide the legal base for air pollution abatement. The knowledge base for air chemistry and pollutant effects was primitive at the time of first enactment. As new knowledge developed, the law was amended in major ways in 1967, 1970, and 1977, but the basic regulatory framework remains the same. The major purpose of the Clean Air Act is "to protect and enhance the quality of the nation's air resources so as to promote the public health and welfare and the productive capacity of its population." [42 USC §7401(*b*)(1)]

In order to achieve its broad purpose, the Clean Air Act, as amended in 1970, established two kinds of national ambient air quality standards. Primary ambient air quality standards are those required to protect human health with an adequate margin of safety. Secondary ambient air quality standards specify the pollutant concentration levels required to protect public welfare from any known or anticipated adverse effects associated with the presence of air pollutants. The concept of public welfare is very broad; secondary standards are based on damage to a variety of resources including crops, vegetation, wildlife, and visibility, as well as adverse effects on the economy *and* climate. An air quality standard is a level to which a pollutant concentration should be reduced to avoid undesirable effects.

The identification of pollutants for which standards need to be set has been driven by considerations of their impact on human health. Emphasis has been placed on the effects of single pollutants, though the "criteria documents" prepared by the Environmental Protection Agency attempt to assess health effects that stem from the interaction of pollutants. The U.S. has set primary and secondary standards for ozone, carbon monoxide, nitrogen dioxide, sulfur dioxide, particulate matter, and lead.

The Clean Air Act provides for the means to implement, maintain and enforce national air quality standards. Central to this process are plans drawn up by the states.

State control strategies can employ a variety of measures to attain national air quality standards, including land use control, gasoline rationing, and emission charges.

A primary means of attempting to ensure the achievement of air quality standards has been the institution of emission controls. The framework of the Clean Air Act is based on the assumption that it is possible to work backward from air quality standards based largely on health effects to determine emission standards in a particular region. At present, our understanding of air chemistry and meteorology falls short of providing the scientific basis for translating air quality standards into emission standards. This problem of setting standards has led to endless controversy as to the "best" means of achieving air quality.

Carbon Monoxide and Ozone Nonattainment

The difficulties with attaining air quality standards under the provisions of the Clean Air Act are illustrated by the history of the nonattainment of standards for ozone and carbon monoxide. These two pollutants differ greatly in their sources. Carbon monoxide is a product of incomplete combustion of fossil fuels that is emitted directly into the atmosphere. As has been noted, CO participates in a variety of reactions, including those leading to ozone formation, and provides a major sink for hydroxyl. Unlike carbon monoxide, ozone has no direct sources. Regulation of ozone can only be accomplished by controlling the emissions of its precursors—volatile hydrocarbons and nitrogen oxides. The Clean Air Act Amendments of 1970 provided up to six years for the attainment of CO and O_3 ambient standards. When the final deadline for attainment passed in 1976 and numerous areas of the country had failed to comply, Congress, in the 1977 Amendments, extended the deadline to 1982. For areas with more severe ozone problems, deadlines were extended until 1987. At the time, it was assumed that, at worst, only the Los Angeles area might fail to meet the standard by 1987. Instead, the 1987 deadline found some 67 areas, with a total population of about 80 million, failing to meet the ozone standard; 88 areas with a 65-million population had missed the CO standard.

The widespread failure to achieve the ambient standards may have been due to poor implementation and enforcement of control programs by national and state authorities, or to the setting of unachievable standards resulting from inadequate understanding of the science, or to some combination of these. The assumption that it is possible to proceed from ambient standards to emission standards is certainly part of the problem. Perhaps as important is the failure to recognize the linkages among pollutants in setting and implementing standards. From the standpoint of their direct effects or their interactivity, pollutants cannot be considered singly; e.g., carbon monoxide presents not only a health hazard but is a participant in the production of ozone, and emphasis on NO_x reduction may actually have led to an increase in ozone formation in some regions. Present legislated restrictions forbid an adequate exploration of the large number of alternative approaches to air pollution control.

Proposed Acid Rain Controls

The Clean Air Act, as formulated in 1963 and amended in subsequent years through 1977, was aimed at protecting human health. The goals of achieving ambient standards were clearly secondary. During the 1980s, when air pollution was implicated in regional acidification of lakes and in forest damage, emphasis was shifted to effects other than those on human health. Popular concerns over "acid rain" led to the identification of sulfur dioxide as a major contributor to secondary damage in both Canada and the U.S. Legislation introduced in the 100th Congress (1987-1988) to amend the Clean Air Act was focused on reducing emission of SO_2 by using flue gas desulfurization units known as scrubbers. This proposal was based on the assumption that the amount of acid deposition was linearly related to the amount of sulfur dioxide placed in the atmosphere. In the end, Congress deferred a decision on further SO_2 reductions, largely because of the enormous cost, but other considerations played a role. As discussed above, forest damage may be due in large part to causes other than acid deposition. The evidence supporting a linear relation between emissions and deposition is contradicted by the seasonal patterns in acid deposition illustrated in Figure 2, and remains a contentious issue.

Europe has moved faster than the United States in instituting controls on SO_2 in order to suppress forest damage. Under the auspices of the Economic Commission for Europe, a number of European countries in 1985 adopted a protocol that, in principle, called for a 30% reduction from 1980 SO_2 emission levels by 1993. However, Great Britain and the United States have refused to sign the protocol, citing scientific uncertainty with respect to cause and effect.

The greenhouse implications of regulatory proposals for acid rain control were not evaluated or even considered in the deliberations of the 100th Congress. The major element of acid rain reduction proposals was the use of flue gas scrubbers to remove SO_2 – a strategy that has both direct and indirect effects on greenhouse warming. Use of scrubbers on coal-fired power plants reduces unit efficiency by the consumption of energy during scrubber operation. With scrubbers, a 6% reduction in the efficiency of converting thermal to electrical energy (35%-33%) appears a conservative estimate (Margolies et al., 1988). The reduction in efficiency leads to a 6% increase in fuel use, with a corresponding increase in carbon dioxide release. Furthermore, eliminating sulfur dioxide from emissions reduces the sink for ozone and other oxidants. Because oxidants are implicated in damage associated with acid rain, the reduction of SO_2 may actually lead to increased forest damage. Evaluation of the net damage associated with SO_2 reduction requires a far greater understanding of atmospheric chemistry and effects than is currently available.

Global Limits on Chlorofluorocarbons

The first step toward limiting global pollutant emissions was taken in 1972 at the United Nations Conference on the Environment in Stockholm. The conference called for an examination of stratospheric pollutants.

Scientific evidence developed during the 1970s and 1980s made it clear that chlorofluorocarbons, which have an atmospheric lifetime of a hundred years or more, were depleting stratospheric ozone.

This evidence prompted the United States, along with Canada and the Scandinavian countries, to voluntarily end the use of CFCs in aerosol sprays in the late 1970s. On September 17, 1987, twenty-four nations plus the European Community signed a protocol agreeing to limits on CFC production, while another twenty-five nations signed documents approving actions taken at the meeting. The major provisions of the protocol are:

1. In 1989, when it takes effect, participating nations are to freeze the use of chlorofluorocarbons at 1986 levels;
2. By 1994, consumption must be reduced by 20%;
3. By 1999, consumption is to be cut by an additional 30%, thus reducing the 1986 consumption level by one-half;
4. Use of halons, chemicals used as fire suppressants, is to be frozen at 1986 levels by 1994, but reductions would not be required.

The Montreal Protocol was the first global agreement to specifically limit use of a pollutant. By August 1988, thirty-five nations had signed the protocol and six countries had ratified.

The limitation of CFCs was triggered by the increase in surface ultraviolet flux that resulted from the destruction these chemicals caused to stratospheric ozone. As in the case of the Clean Air Act, the prime consideration was the protection of human health. The impact of CFCs on global climate, or the repercussions of increased surface ultraviolet on urban smog, played little or no role in considerations leading to the protocol. In fact, the effects CFCs have on the greenhouse are complex, and their net effect is not well determined. Chlorofluorocarbons are extremely strong infrared absorbers on a per-molecule basis, but the chlorine they release on dissociation in the stratosphere starts a chain reaction that destroys hundreds of ozone molecules for each molecule of CFC. Because ozone is a greenhouse gas, the net stratospheric greenhouse balance depends on the detailed chemistry. As previously discussed, a higher surface ultraviolet flux produces tropospheric ozone, adding to the greenhouse burden. The net greenhouse effect of CFCs, though uncertain, appears to be positive (Ramanathan, 1988).

Alternative Control Mechanisms

The Clean Air Act as amended—with its standard-setting and provision for enforcement—has provided the basis for the United States to address the issues connected with a changing atmosphere. U.S. strategy thus far has had major successes in lowering the levels of hazardous pollutants, such as lead, in the atmosphere. However, the regulatory framework has not adequately dealt with the pervasive urban problem of high levels of health-endangering carbon monoxide and ozone. The need to maintain regional air quality and prevent adverse global climate change has introduced complexities not recognized at the time the Clean Air Act was formulated. Advances in science have further complicated regulatory approaches; for example, the exact importance of hydroxyl in atmospheric chemistry (Levy, 1971) was recognized only after the writing of the far-reaching Clean Air Act Amendments of 1970. The numerous linkages among pollutants, and among effects and pollutants, further suggest that other approaches to achieving cleaner air should be probed.

Imposing a financial charge against the emission of species that harmfully alter the atmosphere is a promising approach to improving air quality (Anderson et al., 1977). A well-constructed charge system based on amounts of emission could quickly curb discharge by significantly changing the basic costs facing individual polluters. Pollution charges would provide a strong abatement incentive and would tie environmental costs to the processes that generate offending species. As the cost of pollution control is incorporated into the market structure, producers and consumers of high-pollution goods would carry their share of the burden. Because the price of goods reflects the relative cost of reducing emissions, consumers would, to some extent, shift to goods that embody lower environmental costs.

Advocates who favor either regulatory authority or user fees often consider the two approaches as mutually exclusive; this need not be so. Regulatory approaches can be used, as is the case in the Clean Air Act, to set ambient standards and back-up enforcement, while discharge fees provide the economic incentive to achieve these standards.

In order for user fees to be economically efficient, emission amounts must be determinable without monitoring. If monitoring is required, economic costs for instrumentation, inspection, and enforcement become overwhelming. Of the pollutants discussed in this paper, only two meet the criteria for economically efficient application of user charges: carbon dioxide and sulfur dioxide. Carbon dioxide emissions are fixed by the carbon content of a particular fuel; the small fraction of carbon monoxide that also is generated eventually converts to carbon dioxide in the environment. Emissions of sulfur dioxide are fixed by the sulfur content of the fuel or, in the case of smelters, the sulfur content of the ore. Sulfur removed after combustion by scrubbers or bag houses can be readily determined by tonnages; as in the case of CO_2, monitoring effluents is not required. In the case of fuels, charges would be assessed on either the producer or importer.

The use of charges to help control the flow of emissions into the atmosphere has not been tried in the United States, even though the Nixon administration proposed a nationwide sulfur tax in 1970, and again in 1971. These proposals received a great deal of attention but were not adopted, primarily because of the strong Congressional view that regulation and enforcement were the best means of achieving air quality.

The proposed sulfur charges were designed to create an immediate and continuing incentive to abate sulfur emissions. Charge levels were designed to approximate control costs, which were estimated to equal social damage. The tax was to be 15 cents per pound in regions that had not attained national primary air standards for sulfur dioxide, 10 cents per pound where primary, but not secondary, standards had been met, and no charge in regions meeting secondary standards. In the early 1970s, when these proposals were made, it was not known that SO_2 could undergo long-range transport to be deposited as acid rain. Further, the zero charge in areas that met both primary and secondary standards would have provided an incentive for sulfur emitters to move to those regions. If a charge were to be considered currently, both the impact of long-range transport and new developments in abatement technology would have to be considered.

A charge on carbon dioxide would encourage a shift to those fuels that produce little or no CO_2 per unit of energy delivered: natural gas, nuclear, and solar. Such a charge would provide a strong incentive for energy efficiency in all hydrocarbon fuel use. The charge could be coupled with efficiency standards. Regulated efficiency standards for automobiles, appliances, etc. could furnish a minimum limit to energy efficiency, with the carbon dioxide charge providing additional economic incentives. I have made a preliminary suggestion of one cent per kilogram of carbon placed in the atmosphere (Testimony before the Committee on Energy and Natural Resources, United States Senate, November 10, 1987). At this level, the charge for burning 1,000 cubic feet of natural gas would be 14 cents; the average charge for burning a barrel of oil would be \$1.20, and the fee for burning a short ton of average coal would be \$5.20. However, many difficult issues require resolution: timing the charge imposition to assure energy efficiency gains without placing U.S. industry at a competitive disadvantage, appropriate charge levels, etc.

Reconciliation of Control Measures

Attempting to slow atmospheric degradation solely by means of regulation is an approach for which alternatives need to be sought. In addition, steps to improve urban air quality should be harmonized with the goals of reducing both the regional damage associated with air pollution and the potential for climate change. At the very least, any proposal for change in the current regulatory regime should be accompanied by a detailed consideration of its impact on the acid rain and greenhouse problems. I have discussed the possible climate implications of the use of scrubbers to reduce sulfur dioxide emissions. Another example is given by the proposal that methanol be required as an alternative fuel to gasoline in those regions having a severe carbon monoxide problem (e.g., Denver). The fuel cycle for the production and burning of methanol from natural gas produces about 10% more carbon dioxide than does gasoline, or 50% more CO_2 than does natural gas, in delivering the same automotive mileage. If methanol is derived from coal, its contribution to the carbon dioxide burden of the atmosphere doubles that of gasoline. Proposals for the use of methanol disregard the potential for worsening the greenhouse effect.

As has been repeatedly illustrated in this paper, photochemical oxidants play a key role in atmospheric chemistry, ecosystem damage, and climate. The Environmental Protection Agency has chosen to deal with oxidants by selecting ozone as a proxy for all oxidants. Even with respect to human health effects, it is not clear that ozone is the sole, or even the dominant damaging agent. The discovery of high concentrations of hydrogen peroxide in fine droplets indicates that this compound may add significantly to the damage associated with ozone. In terms of secondary standards, where damage to vegetation, materials, etc. must be taken into account, it is clear that hydrogen peroxide plays an important role. The standards set by EPA for the control of oxidants may have to be refined to include limits for hydrogen peroxides, hydroperoxyl, and nitrate.

At present, limits on carbon monoxide are set solely by considerations of human health impacts. Yet carbon monoxide plays a role in producing ozone and hydrogen peroxide, both potential health hazards, as well as agents known to damage ecosystems. These and other such linkages need to be carefully considered in setting both primary and secondary ambient air standards.

Summary

Rapid progress in atmospheric science and in ecology has revealed a myriad of links among pollutants and effects. The regulatory strategies presently used to deal with the stresses of a changing atmosphere were developed at a time when our understanding of these linkages was rudimentary. New scientific findings have not yet been incorporated into our air pollution control regimen, and the legislative process is sufficiently cumbersome that it is difficult to revise laws and regulations in a timely fashion within the framework of the Clean Air Act.

Two recent scientific advances are of particular significance to control strategies: the recognition that oxidants play a complex role in both atmospheric chemistry and in effects, and the importance of the greenhouse. Current regulatory practice focuses on ozone, but other oxidants such as hydrogen peroxide also play key interactive roles. The effects of oxidants are not limited to endangering human health. Oxidants are required to convert sulfur dioxide and oxides of nitrogen into strong acids; they may also be major contributors to ecosystem damage associated with air pollution. Increased ultraviolet flux resulting from ozone reduction by CFC reactions in the stratosphere will promote the formation of ground-level oxidants – decreasing both local and regional air quality.

The linkages of the greenhouse effect with other air quality issues are pervasive. Increases in surface temperature will increase rates of species reactions, thus promoting the formation of oxidants. Climate change stresses ecosystems, making them more vulnerable to stresses associated with conventional pollutants. Health-damaging ozone is also an important greenhouse species, and methane – another greenhouse contributor – plays an important role in oxidant chemistry.

The newly discovered complexities of atmospheric change and its effects suggest that a re-examination of the fundamental tenets of the Clean Air Act may be required. The historical and well justified emphasis on human health and the emphasis on effects of single pollutants both tend to discount the importance of combined and synergistic effects. The complex interplay between the local, regional, and global consequences of a changing atmosphere cannot be dealt with adequately by means of instruments designed to deal with local smog. At the very least, the implication of treating one problem as a substitute for other problems needs to be explicitly dealt with. Alternatives to purely regulatory strategies, in particular, the use of charges, require re-examination. In the end, the solution to complex problems may require complex solutions involving a variety of mechanisms – all of which need to be coordinated.

References

Anderson, F., A. Kneese, P. Reed, S. Taylor and R. Stevenson (1977), *Environmental Improvement Through Economic Incentives*, J. Hopkins Press: Baltimore.

Bormann, F. (1982), Air pollution stress and energy policy, in *New England Prospects: Critical Choice in a Time of Change* (ed., C. Reidel), UniversityPress : New England, Hanover.

Caldwell, M. (1979), Plant life and ultraviolet radiation: Some perspective in the history of the earth's UV climate, *BioScience*, **29**, 520-525.

Chappelka, A., B. Chevone and T. Bank (1985), Growth response of yellow poplar (Liriodendron tulipfera) seedlings to ozone, sulfur dioxide and simulated acidic precipitation, alone and in combination, *Environ. Exp. Biol.*, **25**, 233-244.

Crutzen, P. (1987), Role of the tropics in atmospheric chemistry, in *The Geophysiology of Amagonia* (ed., R. Dickenson), Wiley: New York, 107-130.

Crutzen, P. (1988), Tropospheric ozone: An overview, in *Tropospheric Ozone* (ed., I. Isaksen), Reidel: Dordrecht.

Farman, J., B. Gardiner and J. Shanklin (1985), Large losses of total ozone in Antarctica reveal seasonal ClO_x/NO_x interaction, *Nature*, **315**, 207-210.

Finlayson-Pitts, B. and J. Pitts (1986), *Atmospheric Chemistry: Fundamentals and Experimental Techniques*, Wiley: New York.

Fishman, J., and W. Seiler (1983), Correlative nature of ozone and carbon monoxide in the troposphere: Implications for the tropospheric ozone budget, *J. Geophys. Res.*, **88**, 3662-3670.

From, E. and C. Keeling (1986), Reassessment of late 19th century atmospheric carbon dioxide variations in the air of western Europe and the British Isles based on unpublished analysis of contemporary air masses by G. S. Callendar, *Tellus*, **38B**, 87-105.

Gamlen, P., B. Lane, P. Midgley and J. Steed (1986), The production and release to the atmosphere of CCl_3F and CCl_2F_2 (chlorofluorocarbons CFC 11 and CFC 12), *Atmos. Env.*, **20**, 1077-1085.

Haagen-Smit, A. (1952), Chemistry and physiology of Los Angeles Smog, *Indust. Eng. Chem.*, **44**, 1342 -1346.

Haagen-Smit, A., C. Bradley and M. Fox (1952), Formation of smog in Los Angeles smog, *Proc. Second Nat'l. Air Pollut. Symp.*, 54-56.

Heikes, B., G. Kok, J. Walega and A. Lazrus (1987), H_2O_2, O_3 and SO_2 measurements in the lower troposphere over the eastern United States during fall, *J. Geophys. Res.*, **92**, 915-931.

Henderson, R. and K. Weingartner (1980), Analysis of MAP3S precipitation chemistry data-implications for modeling, MITRE Technical Report (MTR-80W00265), The MITRE Corporation: McLean, Va.

Levy, H. (1971), Normal atmosphere: Large radical and formaldehyde concentrations predicted, *Science*, **173**, 141-143.

Liu, S., M. Trainer, F. Fehsenfeld, D. Parrish, E. Williams, D. Fahey, G. Hübler and P. Murphy (1987), Ozone production in the rural troposphere and the implications for regional and global ozone distributions, *J. Geophys. Res.*, **92**, 4191-4207.

Margolies, A., T. Vivenzio, B. Feldman and E. Zabolotny (1988), Potential application of slagging combustion systems, in *Proc. Energy Tech. Conf.* (held in Washington, D.C. February 17-19, 1988), Government Institutes: Rockville, Md.

Masuch, G., A. Kettrup, R. Mallant and J. Slanina (1986), Effects of H_2O_2-containing acidic fog on young trees, *Inter. J. Environ. Anal. Chem.*, **27**, 183-213.

McLaughlin, S. (1985), Effects of air pollution on forests: A critical review, *J. Air Pollut. Control Assoc.*, **35**, 512-534.

Meagher, J., K. Olszyna and M. Luria (1984), The effect of SO_2 gas phase oxidation on hydroxyl smog chemistry, *Atmos. Env.*, **18**, 2095-2104.

Miller, P., J. Parmeter, B. Flick and C. Martinez (1969), Ozone dosage response of Ponderosa pine seedlings, *J. Air Pollut. Cont. Assoc.*, **19**, 435-538.

Munger, J., D. Jacob, J. Waldman and M. Hoffmann (1983), Fogwater chemistry in an urban atmosphere, *J. Geophys. Res.*, **88**, 5109-5121.

Perner, D., U. Platt, M. Trainer, G. Hubler, J. Drummond, W. Jurkerman, J. Rudolph, B. Schubert, A. Valtz and D. Enbalt (1987), Measurement of tropospheric OH concentrations: A comparison of field data with model prediction, *J. Atmos. Chem.* **5**, 185-216.

Pittock, A. and H. Nix (1986), The effect of changing climate on Australian biomass production - a preliminary study, *Climate Change*, **8**, 243-255.

Ramanathan, V. (1988), The greenhouse theory of climate change: A test by an inadvertent global experiment, *Science*, **240**, 293-299.

Ramanathan, V., R. Cicerone, H. Singh and J. Kiehl (1985), Trace gas trends and their potential role in climate change, *J. Geophys. Res.*, **90**, 5547-5566.

Reich, P. (1987), Quantifying plant response to ozone: A unifying theory, *Tree Physiol.*, **3**, 63-91.

Seinfeld, J. (1986), *Atmospheric Chemistry and Physics of Air Pollution*, Wiley: New York.

Shugart, H., M. Antonovsky, P. Jarvis and A. Sandford (1986), CO_2, climate change and forest ecosystems, in *The Greenhouse Effect, Climate Change and Ecosystems* (ed., B. Bolin, D. Döös, J. Jager and R. Warrick), Wiley: New York.

Stolarski, R., A. Krueger, M. Schoeberl, R. McPeters, P. Newman and J. Alpert (1986), Nimbus 7 satellite measurements of the springtime Antarctic ozone decrease, *Nature*, **322**, 808-811.

Wieler, A. (1911), *Pflanzenwachstum und Kolpmanzel im Boden*, Bornträfer: Berlin.

GLOBAL ENERGY USE: A QUANTITATIVE ANALYSIS

Richard Wilson

Department of Physics, and Energy and Environmental Policy Center
Harvard University
Cambridge, MA 02138

Abstract

The historical record of fuel supply and energy demand until 1973 is generally agreed. Since 1973, there have been many projections and it is possible to examine the record between 1973 to 1988 to understand their validity. The old idea that the ratio of GNP/energy is constant was discredited early; however, GNP is more closely coupled to electricity growth. This is an indication of a fact derivable on other grounds; measures to achieve fuel energy efficiency apply mostly to the non-electrical sector, and can *increase* the consumption in the electricity sector. The historical record shows a fundamental difference in purpose between attempts to keep the cost of energy low, and creation of incentives to reduce fossil fuel burning. The role of nuclear power is presently limited to the electrical sector. There it is interchangeable with coal and can, given reasonable construction times, replace coal approximately independently of any other fuel reductions. Whether or not society wishes to do so is a question independent of the efficiency of fuel use.

Introduction: Society's Use of Energy

Whenever I consider global energy use, I find it useful to consider how man's use of fuels has changed over the centuries. Each major change has the aspects of a revolution.

Primitive man used only the food he could eat – about 100 watts, the size of a large light bulb. This limited the tasks that he could undertake, and the force he could bring to bear for these tasks. Several thousand years ago man domesticated animals. This gave him access to some fuels that he otherwise could not use, such as grasses. This also enabled him to use a force 10 times as great as before. Using horses and cattle, man controlled as much as 1,000 watts per person – ten times as much as he ate, and was able to exert a force ten times as large as previously.

Copyright 1989 by Elsevier Science Publishing Company, Inc.
Global Climate Change Linkages
James C. White, Editor

About 2,000 years ago, man developed waterwheels and soon thereafter he developed windmills. These developed 10,000 watts 1,000 years ago, and up to a million watts now. These inventions were so efficient, for the time, that it was economic to send wheat to the miller to be ground rather than grinding it individually. This accelerated a concentration of activity in villages, where the mills could be used for a whole community.

The next major change came about 250 years ago as the steam engine was developed. For example, using Newcomen's new steam engine, the drydock of Catherine the Great at St. Petersburg (now Leningrad) was emptied in eight hours, whereas previously it had taken 20 days by windmill to empty it. This harnessing of the steam engine was particularly important because the engine could be put on a wagon and be self-propelled. This fantastic invention led to the development of railroads. Railroads made an enormous contribution to the development of both the American West and Soviet Siberia.

This invention, which brought men closer to each other and made them more interdependent, was soon followed by another that enabled him to "get away from it all" and achieve a certain measure of independence. This was the discovery of oil, and the invention of the internal combustion engine to use it. Now man had a simple, reliable small engine. With Henry Ford's genius added to this invention, each man became independently mobile.

The most recent revolution was the discovery of nuclear fission and the construction of devices to use fission energy. This gives us more fuels – with a breeder reactor an almost unlimited supply – but with an energy density three million times what was previously available. As we all know, this has already made major changes. We now know how to make a bomb three million times as powerful as before – with consequences that we all know. We also have the possibility of reducing the environmental consequences of fuel use. We can reduce the amount of mining that is necessary 300 times if we use the present generation of reactors which only burn the rare isotope uranium 235, and 30,000 times when we develop breeder reactors that utilize the uranium 238 as well. We can purify fuels before burning and purify them after burning. We cannot do this with fossil fuels. We can contemplate keeping wastes out of the environment for thousands of years, which we cannot contemplate with any other waste product of society.

Finally today, we now realize that man is having such a large effect in the environment that we are approaching the limits of the ability of the environment to absorb the effluents, particularly of fossil fuel burning. For some hundreds of years this has happened on a local or regional scale (the London fogs of my youth being an example) but now this is happening globally. Accordingly, we must consider the effect on the global environment of each and every fuel source and every use of energy.

Global Environmental Effects

We have long known that the earth is a greenhouse. Fourier suggested it 150 years ago, and 90 years ago Arrhenius (1896) made the first quantitative calculations of the effects. Arrhenius suggested that doubling of CO_2 concentration would lead to a 3°C global warming. This is the center of the presently accepted range. The earth's atmosphere holds in much of the infrared radiation that it emits. Several gases contribute to the outer atmosphere layer that holds in the gases: carbon dioxide (CO_2), methane (CH_4), fluorocarbons, and nitrous oxide (N_2O). The concentrations of all of these are rising. Presumably these are holding in more infrared radiation than before. The increase in concentrations has serious potential consequences and we are here considering how we can reduce these emissions. If worldwide CO_2 emissions continue at the present rate, the concentrations will double within 200 years. If worldwide CO_2 emissions go up, the doubling time will be less. The "half life" for absorption of CO_2 by deep oceans is close to 1,000 years, so that we might end up with a ten-fold increase in CO_2 concentrations in 1,000 years.

It is also noteworthy that methane and nitrous oxide also are related to energy use. Atom for atom, they are more important than CO_2. However, CO_2 is responsible for about half of the total greenhouse absorption.

The poverty and impotence of present federal policy is illustrated by the fact that Congress has asked EPA for a scenario to "stabilize" the CO_2 *concentrations*. This is taken by EPA merely as a request for a scenario to keep the *emissions* constant! However, even this is hard as we shall see.

Historical Energy Use

My assignment is to discuss global energy use and projections that are made of future use; the purpose is to help in consideration of what can and should be done to reduce emissions of CO_2. It is, of course, part of this assignment to consider especially those fuel uses that lead to the carbon dioxide problem. I start this assignment by going back to the beginning. In Figure 1, I show the total per capita primary energy use in different parts of the world in 1982. U.S. and Canadian citizens, now 250 million of us, used an average of 11.4 kw of energy. This is equivalent to each of us using continuously four one-kw electric fires with the power station rejecting as waste heat the remaining 7.4 kw. In other developed nations, 923 million people use 4.4 kw, and in central Africa and southeast Asia, 1,422 million people use 0.2 kw – barely more than their food intake. The 1,364 million in other developing countries use a little more. This is also shown in Table I taken from the World Commission on Environment and Development (WCED, 1985). As we shall see, most of the fuels that the United States uses are fossil fuels – containing carbon and producing carbon dioxide when they are burned.

Our hopes for the future of humankind must include the hope the 2,800 million people in developing countries can raise their standard of living – and this must include more energy use. If we are to keep global energy use down, we must reduce use in the U.S. and developed countries. For this reason, and for the reason that for the U.S. we have good data, I start by discussing U.S. energy use over the last couple of centuries.

I will describe the world energy picture, not with my own graphs, but with those of others. However, I have checked these for general accuracy and consistency. In Figure 2, I show primary energy use, broken down by fuel source, in the U.S., taken from Hottel and Howard (1972). I extend this graph from 1970 to 1987. We see that 200 years ago

Table I

Global Primary Energy Consumption Per Capita, 1984

World Bank GNP Economy Category	GNP Per Capita (1984 dollars)	Energy Consumption (kW per capita)	Mid-1984 Population (million)	Total Consumption (TW)
Low Income	260	0.41	2,390	0.99
Sub-Saharan Africa	210	0.08	258	0.02
Middle Income	1,250	1.07	1,188	1.27
Lower-middle	740	0.57	691	0.39
Upper-middle	1,950	1.76	497	0.87
Sub-Saharan Africa	680	0.25	148	0.04
High-Income Oil Exporters	11,250	5.17	19	0.10
Industrial Market Economies	11,430	7.01	733	5.14
East European Non-Market Economies	--	6.27	389	2.44
World	--	2.11	4,718	9.94

World Commission on Environment and Development
Common Challenges

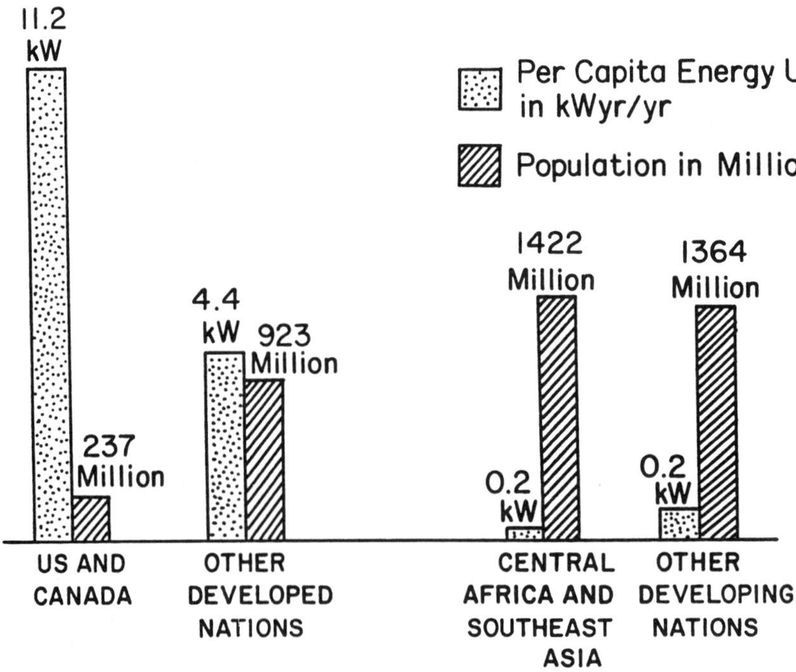

Figure 1

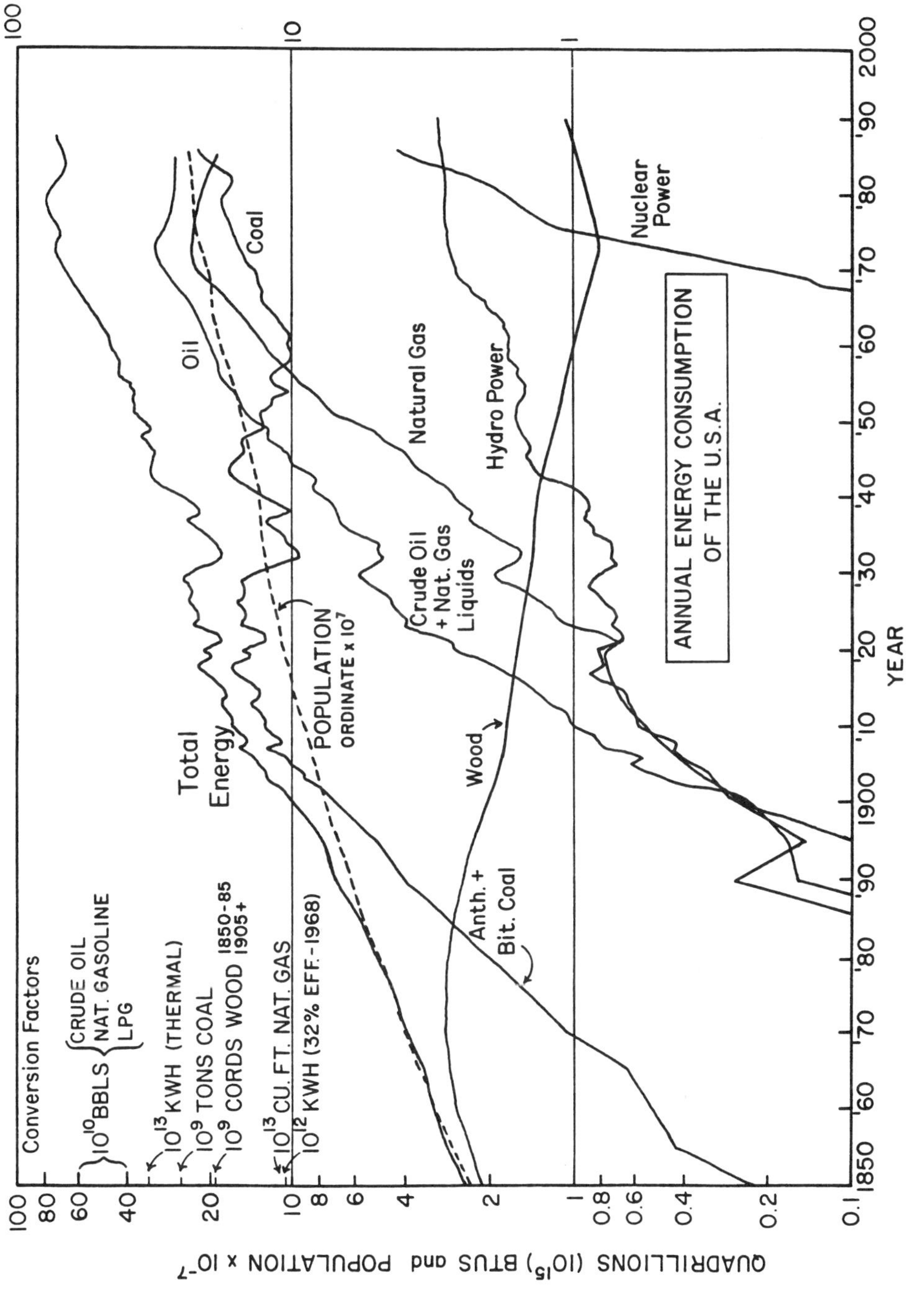
ANNUAL ENERGY CONSUMPTION OF THE U.S.A.
Conversion Factors
10^10 BBLS {CRUDE OIL, NAT. GASOLINE, LPG}
10^13 KWH (THERMAL)
10^9 TONS COAL
10^9 CORDS WOOD 1850-85 1905+
10^13 CU. FT. NAT. GAS
10^12 KWH (32% EFF.-1968)
Total Energy
Oil
Coal
Natural Gas
Hydro Power
Nuclear Power
Crude Oil + Nat. Gas Liquids
POPULATION ORDINATE x 10^7
Wood
Anth. + Bit. Coal
QUADRILLIONS (10^15) BTUS and POPULATION x 10^-7
YEAR
100 80 60 40 20 10 8 6 4 2 1 0.8 0.6 0.4 0.2 0.1
1850 '60 '70 '80 '90 1900 '10 '20 '30 '40 '50 '60 '70 '80 '90 2000
Figure 2

we had what we might now call a "solar economy." We used "passive solar" to reduce demand, and biomass (wood) to actively supply fuel. Even the early railroad engines in the U.S. used wood, unlike their European counterparts that used coal from the inception of railroads. 130 years ago we switched from wood to coal and shortly thereafter to oil. Another "solar" source, which had already contributed as the waterwheel, now began to contribute as hydroelectric power starting at the turn of the century and is still efficient today. Finally the graph shows the rapid rate at which nuclear energy has entered the U.S. economy.

In Figure 3 I show the breakdown of U.S. fuel use both by type of fuel and by sector of society.

I note that total energy use followed approximately the increase in population until 1900, when the easy availability of coal allowed it to surge ahead. We also see that for most of the 11.4 kw/person of energy (70 quads for the country), about 8 quads were resources that do not produce carbon dioxide, 4 quads of nuclear, 3 quads of hydropower and 1 quad of wood.

Of the fuels, it is primarily the fossil fuels, coal, oil and natural gas, that produce carbon dioxide. (The wood actually produces carbon dioxide as it burns, but takes it out of the environment as the trees are grown, and keeps it out until the trees are burned or decay.) Table II shows the relative contribution of each to the CO_2 problem, normalized to a given amount of energy output. Natural gas, CH_4, produces the least CO_2. Hydrogen also combines with oxygen during burning; the hydrogen oxidizes to water which is relatively less important than CO_2. Coal produces nearly twice as much CO_2 per unit of energy output as gas.

In addition, every fuel that is burned in air adds to the nitrogen oxide problem. The burning is more controllable for gas than for a solid and the N_2O production is smaller. Once again coal is the worst. Very roughly, the emission of N_2O on burning adds an extra 20% to the greenhouse warming, but this can be reduced by careful combustion. CH_4 also contributes directly to the greenhouse effect and here each molecule of CH_4 is 20 times as bad as each molecule of CO_2. In extracting CH_4 from the ground (or elsewhere), care must therefore be taken to ensure that there is no leakage.

I note from Figure 2 the dominant use of coal in the world's economy. Legasov (1986) has said that 35% of rail transport in the U.S.S.R. is transporting coal. The number is nearly 40% in the U.S. Figure 4 shows a coal train. In Figure 5 I show a coal barge coming up the Huangze River in Shanhai, China. It had been shipped by sea from Qingdao. (Probably it had previously come by train from Taiyuan in Shanxi to Qingdao.) Even people are major carriers of coal. Figure 6 shows a Chinese carrying coal up Mount Emei in Sichuan.

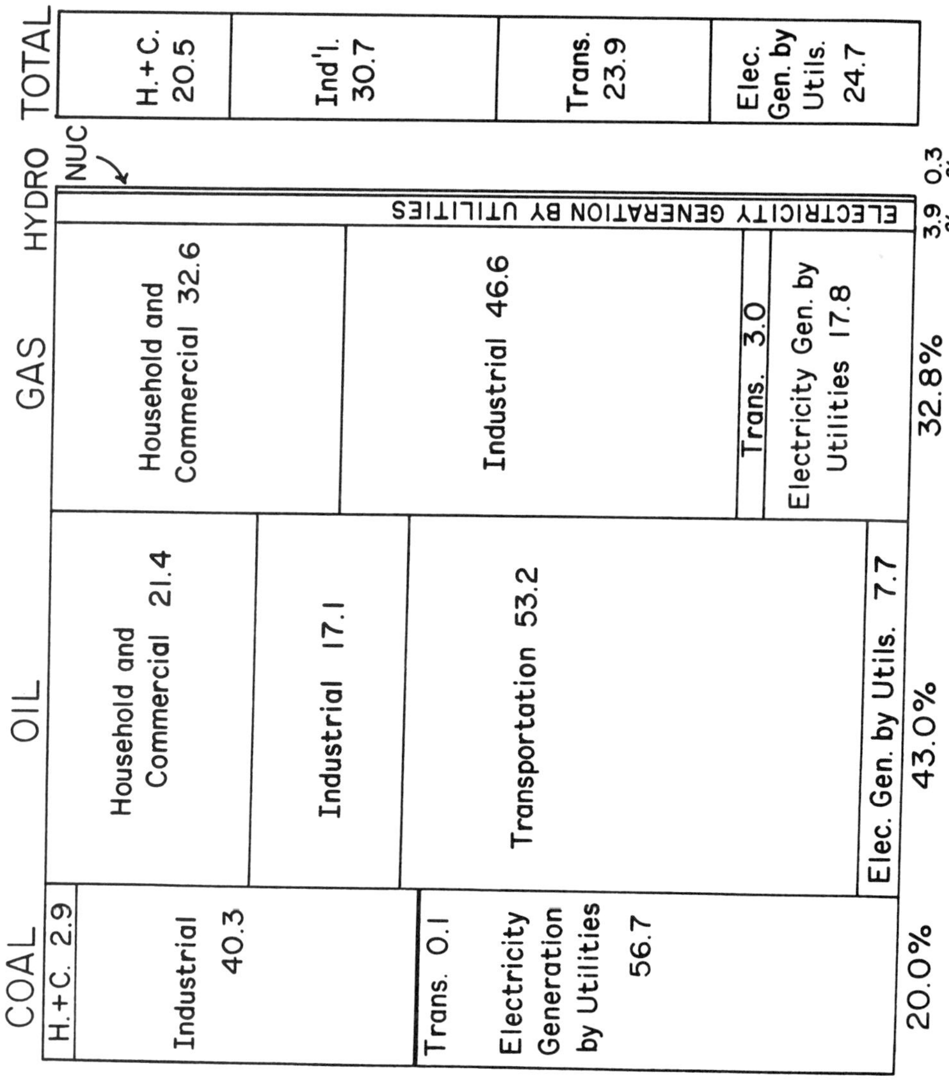
COAL
H.+C. 2.9
Industrial 40.3
Trans. 0.1
Electricity Generation by Utilities 56.7
20.0%
OIL
Household and Commercial 21.4
Industrial 17.1
Transportation 53.2
Elec. Gen. by Utils. 7.7
43.0%
GAS
Household and Commercial 32.6
Industrial 46.6
Trans. 3.0
Electricity Gen. by Utilities 17.8
32.8%
HYDRO
ELECTRICITY GENERATION BY UTILITIES
3.9 %
NUC
0.3 %
TOTAL
H.+C. 20.5
Ind'l. 30.7
Trans. 23.9
Elec. Gen. by Utils. 24.7

Figure 3

Table II

Carbon dioxide emission rates in kg of carbon per million BTU of energy (from Marland, 1982)

Fuel	$kg/10^6$ BTU
Natural gas	14-15
Liquid from crude oil	19-22
Bituminous coal	25
Shale oil	30-110
Liquids from coal	32-54
High BTU gas from coal	34-43

Figure 4

Figure 5

Figure 6

Projections 1970 to 1988: The Energy/GNP Ratio

Projecting energy use into the future is a hazardous business. It has been a necessary tool of the utility industry for years; they will, however, tend to project slightly high, for they have an obligation that society imposes upon them to produce electricity upon demand and a failure to meet this obligation is more serious for them than the consequences of overcapacity. There are, no doubt, some persons who prefer to overestimate potential for energy efficiency.

There are basically two ways to project demand, firstly to understand all about society, including its future inventions and future developments, and secondly to project from the past, assuming that no fundamental change will occur. Before 1973 the second method worked much better than the first. The simple procedure is to plot energy use against time and extrapolate. A typical projection is shown in Figure 7. However, a slight modification worked better still. The extrapolation works better if energy use is plotted against GNP. Then it is assumed that it is desired, and expected, that GNP will slowly increase, and energy will increase with it.

Figure 7

Global Energy Use, 1950-1985*

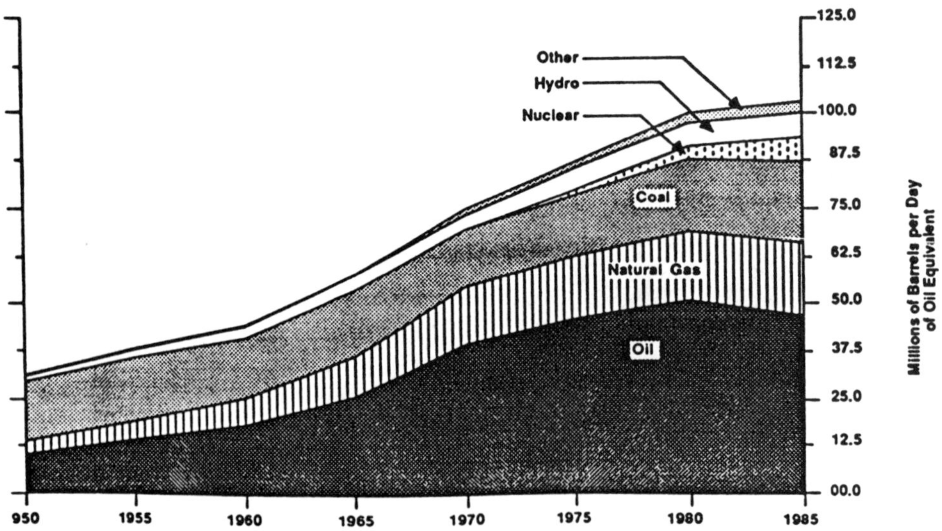

*Excludes centrally planned economies

One of the first Americans to relate fuel supply to economic well-being was Benjamin Franklin who stated:

"In travelling I have observed that in these parts where the inhabitants can have nither [sic] wood nor coal nor turf [peat] but at excessive prices, the working people live in miserable hovels and have nothing comfortable about them. But where fuel is cheap, or where they have the art of using to advantage, they are well furnished with necessaries and have decent habitations."

I note also that he used the subsidiary clause "using fuel to advantage" which we shall see is still very important. I discuss this as "efficient fuel utilization" and others use the phrase "energy conservation."

Figure 8 shows such a graph, made by Chauncey Starr. In 1973 there was, as we all know, a hiatus. After the Yom Kippur war, there was a huge rise in the price of oil, followed by a rise in the price of other primary energy. This price rise led to a major change in societal attitudes. The increase in price, and the almost universal realization that further price increases were likely, set in motion a trend to improve the efficiency with which we use fuels. These made major changes in the structure of the energy economy. I note that in this graph fuel wood and animal feed were *not* included in energy output. If they were, the E/GNP drop would have been steadier.

The historical development of the energy/GNP ratio for the U.S.A. is shown in Figure 9 from 1880 onwards. This figure shows the approximate constancy from the 1946 to 1973 period, but this was superimposed upon a general fall. Between 1945 and 1973, the ratio of energy use to GNP was roughly constant. I note that this was a period of steadily declining oil prices, and hence a period during which there were declines in prices of all forms of energy. Most analysts, during these years before 1973, assumed that the energy/GNP ratio would remain constant during the period of extrapolation. This constancy, combined with an optimistic projection that the GNP would continue to rise steadily, led to predictions of high energy use. But the ratio had not always been constant and the projections were unrealistic.

An energy report from the International Institute for Applied Systems Analysis (IIASA, 1981) listed and projected the energy/GNP ratios for several countries. These lists and projections are shown in Figures 11 and 12. The IIASA analysts noted the historical trend in the U.S. and suggested that the energy /GNP ratio will continue to fall with time and with increasing GNP. Figures 11 and 12 differ primarily in their assumptions of how much GNP will expand, rather than in the plot of energy/GNP vs. GNP.

Several reasons were adduced: increased energy use efficiency, correlated partially with increased GNP itself; and new industries that add to GNP but are less energy intensive than the heavy "smokestack" industries of the past.

However, in developing countries the IIASA analysts believe that a rise in the energy/GNP ratio is still likely, as happened in the U.S. from 1880 to 1910. This is a major respect in which IIASA differs from many advocates of a low energy growth scenario. If IIASA is right, we are holding developing countries back by overemphasizing low energy growth. I also note here the very high ratio of energy/GNP in the U.S.S.R. I attribute this to the particular political and economic systems that they have, that do not seem to provide the proper incentives to become fuel efficient. Since some low energy growth advocates want *restrictions* on the fuel sources they do not like, the effect of these *restrictions* on the general market economy and fuel efficiency must be carefully considered.

I return to the projection of energy use vs. GNP in Figure 8 first made by Chauncey Starr in 1977. He first pointed out, correctly, that increased attention to pollution con-

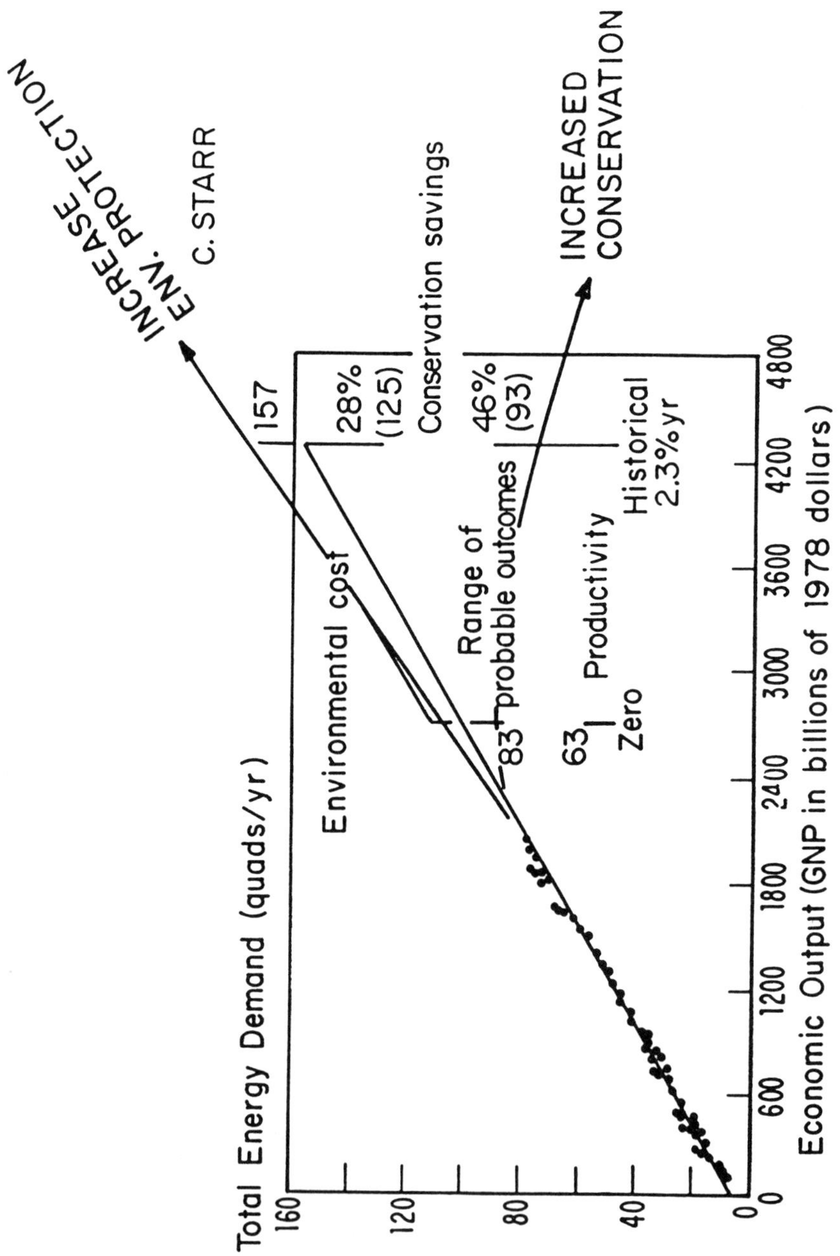

Total Energy Demand (quads/yr)
Economic Output (GNP in billions of 1978 dollars)
0
40
80
120
160
0
600
1200
1800
2400
3000
3600
4200
4800
INCREASE ENV. PROTECTION
C. STARR
Conservation savings
INCREASED CONSERVATION
157
28%
(125)
46%
(93)
Environmental cost
Range of probable outcomes
83
63
Zero
Productivity
Historical 2.3%/yr

Figure 8

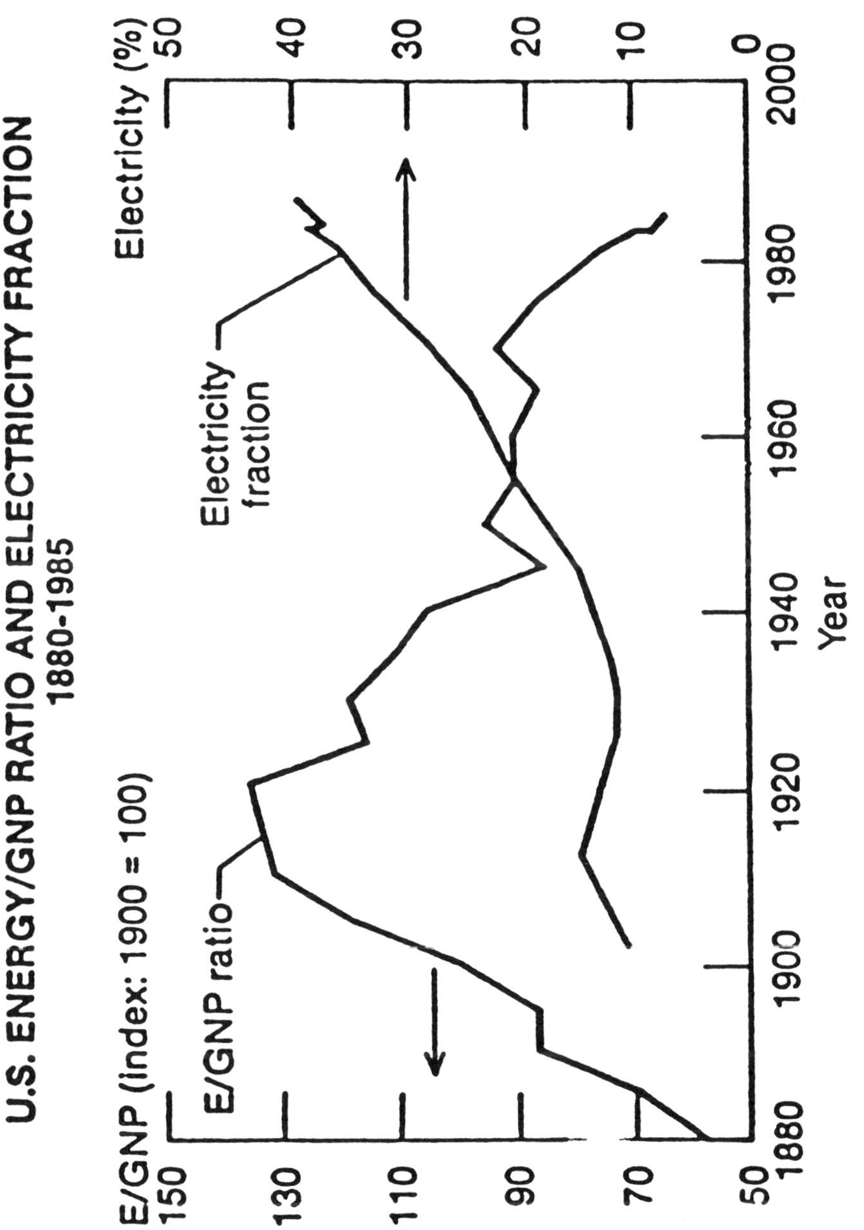
U.S. ENERGY/GNP RATIO AND ELECTRICITY FRACTION
1880-1985
E/GNP (index: 1900 = 100)
Electricity (%)
150
130
110
90
70
50
50
40
30
20
10
0
E/GNP ratio
Electricity fraction
1880
1900
1920
1940
1960
1980
2000
Year

Figure 9

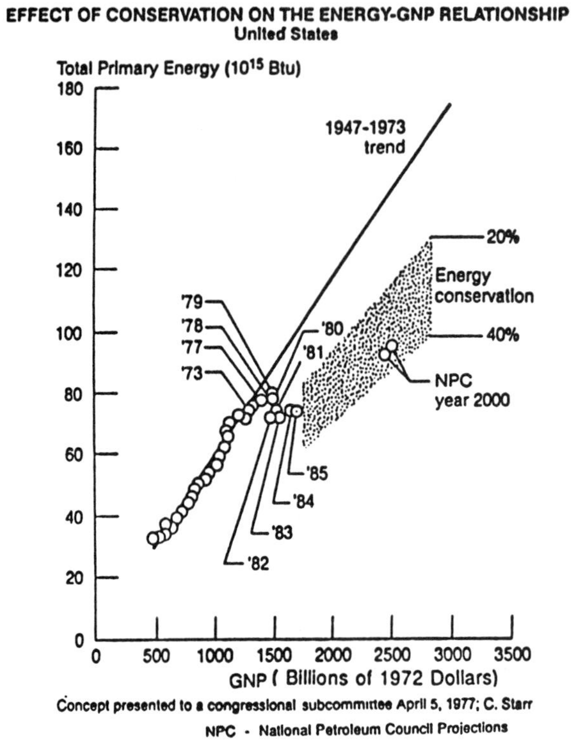

Figure 10

The IIASA world regions.

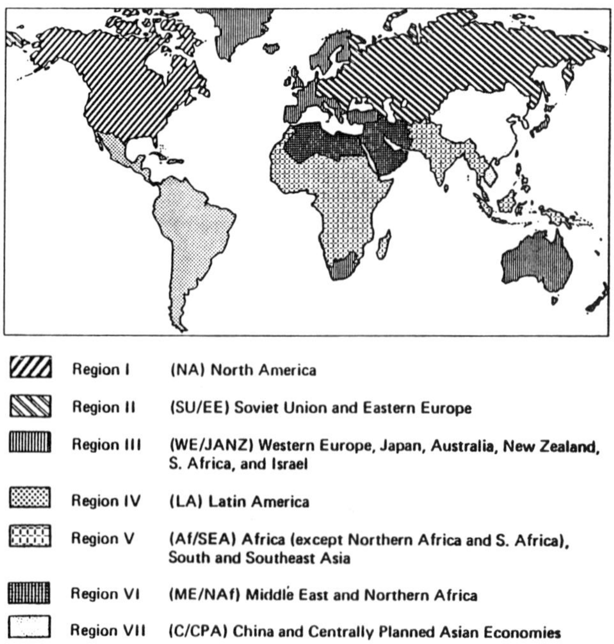

Figure 11

Primary energy and GDP per capita, IIASA regions, 1950-2030.

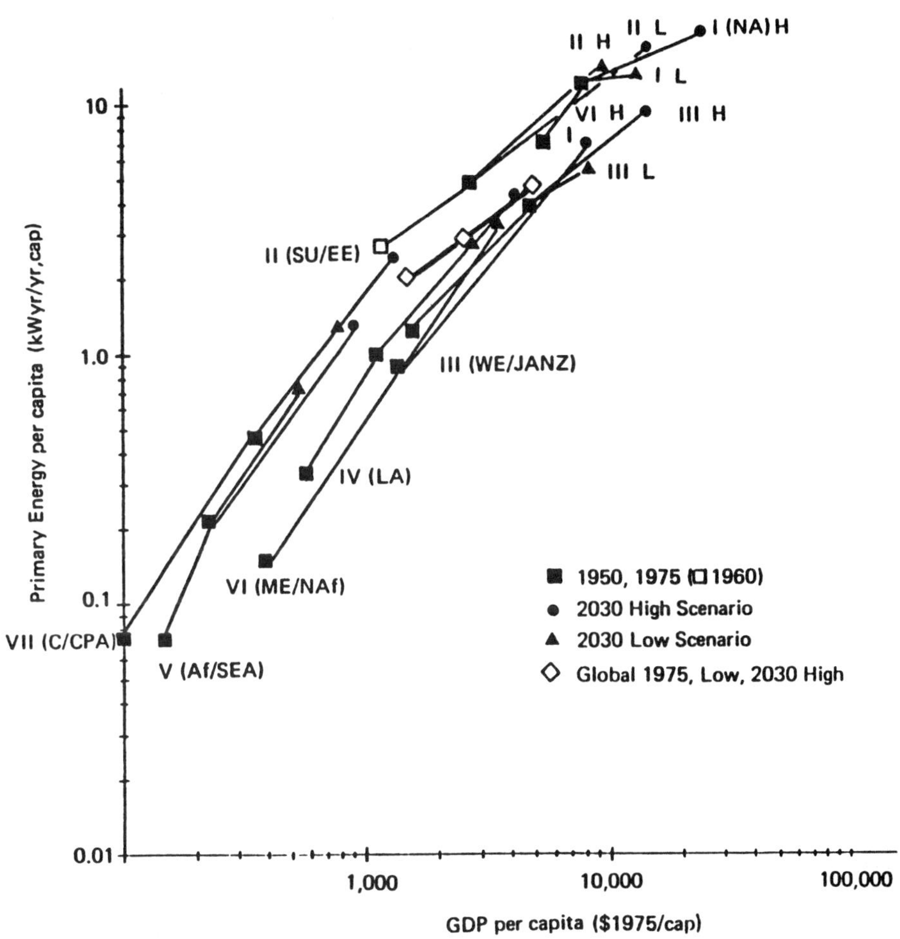

NA = North America
SU/EE = Soviet Union and Eastern Europe
WE/JANZ = Western Europe, Japan, Australia, New Zealand
South Africa and Israel
LA = Latin America
Af/SEA = Africa (except No. Africa and So. Africa,
South and Southeast Asia
ME/NAf = Middle East and Northern Africa
C/CPA = China and Centrally Planned Asian Economies

Figure 12

trol and other environmental improvements would *increase* fuel use if no other steps were taken. But then he assumed that energy efficiency (often called energy conservation) would bring the use below the line. He suggested two possible reductions: 20% at $3,000 billion GNP and 40%. Figure 10 shows almost this same plot (differing primarily in using 1972 dollars) with more recent data superimposed. The *trend* of energy vs. GNP has indeed changed roughly as Starr had suggested. The 1987 fuel use falls on this chart, with the 40% fuel efficiency reduction. However, I note that, although GNP has increased, the rate of increase was not as fast as before, or as projected.

I now want to show that this rapid fall in the ratio of total energy use to GNP has been caused by only *some* energy of the fuels. I show in Figure 13 the distribution of primary energy consumption in the years 1985 to compare with that for 1970 shown in Figure 2. Although the format differs in these pictures, it is clear that oil, which was most affected by the 1973 price rise, was and is used little for electricity generation. One would expect that electricity, which is produced from coal, hydro and nuclear, would be less affected by any change in oil price.

This was looked at by the Committee on Electricity in Economic Growth of the U.S. National Academy of Sciences whose membership is shown in Table III. I show their plot in Figure 14 for 1947 through 1983, *electricity consumption* (not total energy use) versus appropriate indicators: Disposable Personal Income (DPI) for residential use; Gross Product (GPO) for commercial and industrial use. These graphs show a slight break in the top, which is all that is left of the sharp break that the similar graph for total energy shows. In Figure 15 for *world* electricity production even this kink is not seen. Chauncey Starr had already foreseen this (Starr, 1980); this is shown in Figure 16. With more detail he found a break and a decline in electricity vs. GNP but no sudden drop. I note that the points for the years 1977 to 1986 fall directly in the region of the graph that he predicted earlier. There are good reasons why electricity consumption does not follow the same trend as total energy consumption. These are discussed in detail in the National Academy report mentioned above.

These reasons include:

- Industry managers prefer to have the electric utility companies cope with the environmental problems (and regulatory hassles) of energy use.
- Efficient fuel utilization often leads to increased use of electricity. This can, for example, happen as we replace oil heat by an electrically operated heat pump.
- Electricity consuming industries add more to GNP than non-electricity consuming (smokestack) industries.

In Figures 17 and 18, we show the *electricity* use patterns in the residential sector in 1950 to 1980, showing how this increase occurred. Very little electricity is for electric space heat. Yet it is electrical space heat that was often (erroneously) regarded as especially wasteful—almost sinful. In fact, the ease of control of electricity enables it to be used in a fuel efficient way, especially in heat pumps (Burwell, et al., 1983).

Patterns of Energy Use in 1985 for the United States, Western Europe, and Japan

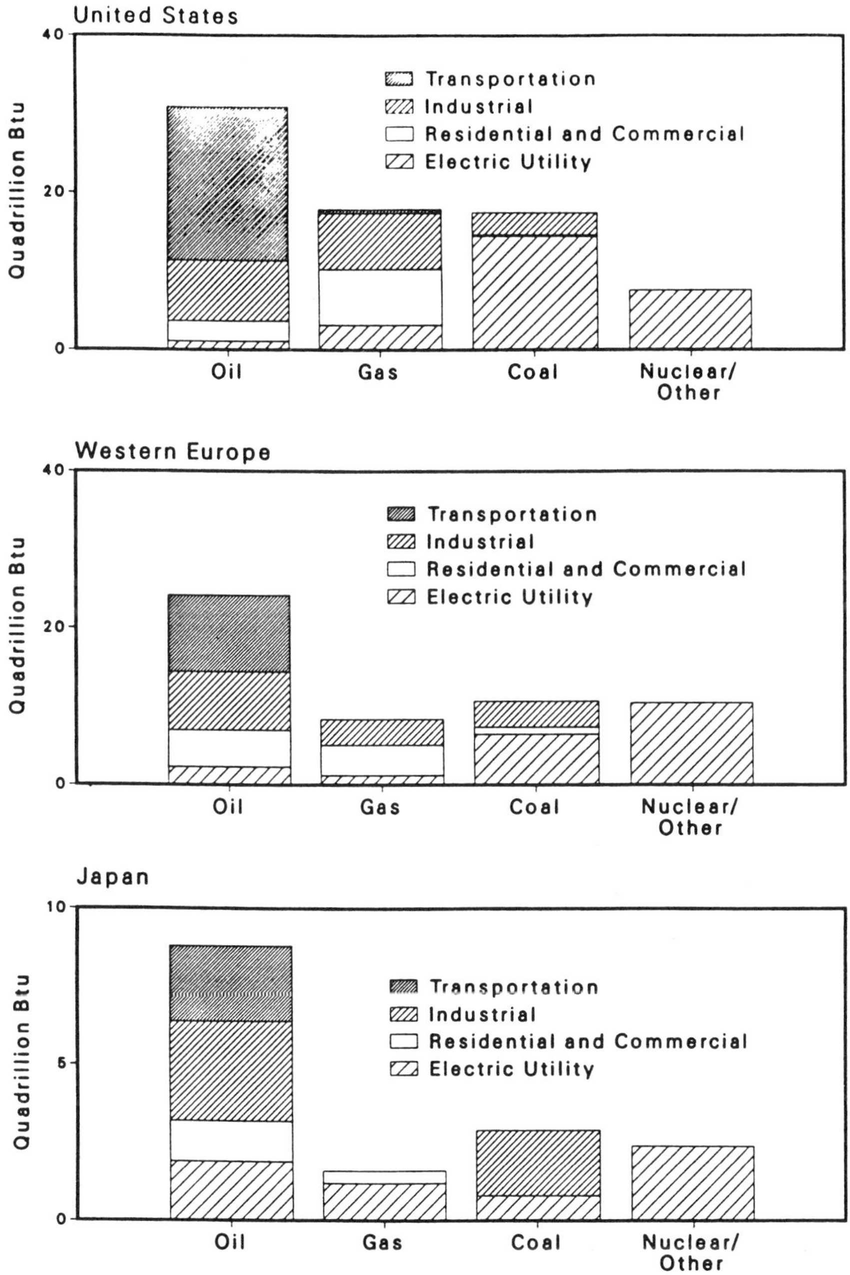

Sources: International Energy Agency, *Energy Balances of OECD Countries, 1970/1985* (Paris, 1987). Energy Information Administration, *Annual Energy Review 1986*, DOE/EIA-0384(86) (Washington, DC, 1987).

Figure 13

Table III

COMMITTEE ON ELECTRICITY IN ECONOMIC GROWTH

U.S. National Academy of Sciences

Martin Baughman (Chairman), Dept. of Electrical and Computer Engineering, The University of Texas at Austin

Barbara R. Barkovich, University of California, Berkeley, California

Atif S. Debs, Dept. of Electrical Engineering, Georgia Institute of Technology, Atlanta, Georgia

Dale W. Jorgenson, Dept. of Economics, Harvard University, Cambridge, Massachusetts

Paul L. Joskow, Dept. of Economics, Massachusetts Institute of Technology, Cambridge, Massachusetts

Eugene W. Meyer, Kidder, Peabody and Company, Inc., New York, New York

Lawrence T. Papay, Southern California Edison Company, Rosemead, California

David B. Roe, Environmental Defense Fund, Berkeley, California

Sam H. Schurr, Electric Power Research Institute, Palo Alto, California

Robert L. Seale, Dept. of Nuclear and Energy Engineering, University of Arizona, Tucson, Arizona

Mason Willrich, Pacific Gas and Electric Company, San Francisco, California

Liaison with Energy Engineering Board: George S. Tolley, Dept. of Economics University of Chicago, Chicago, Illinois

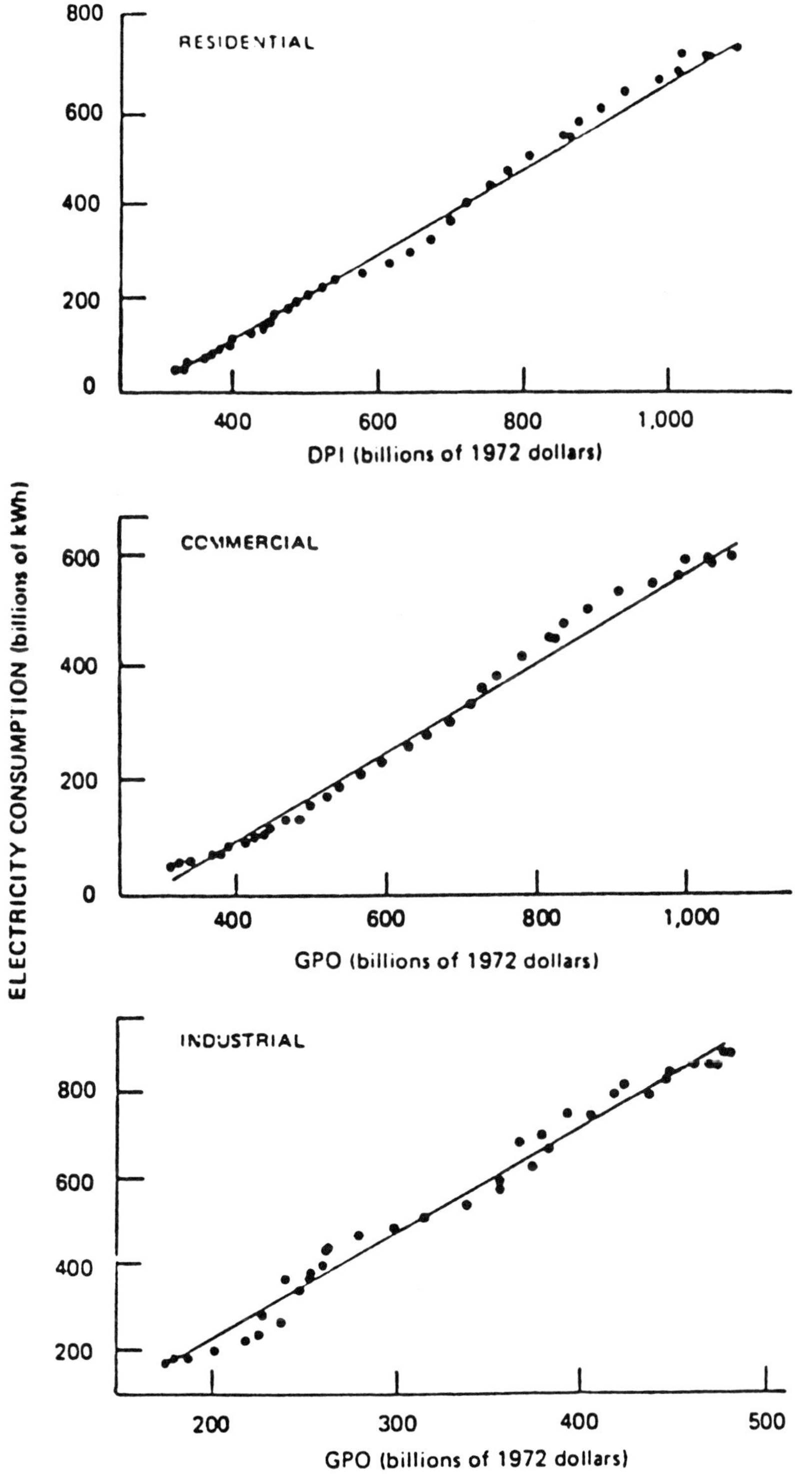
RESIDENTIAL
800
600
400
200
0
400
600
800
1,000
DPI (billions of 1972 dollars)
ELECTRICITY CONSUMPTION (billions of kWh)
COMMERCIAL
600
400
200
0
400
600
800
1,000
GPO (billions of 1972 dollars)
INDUSTRIAL
800
600
400
200
200
300
400
500
GPO (billions of 1972 dollars)

Figure 14

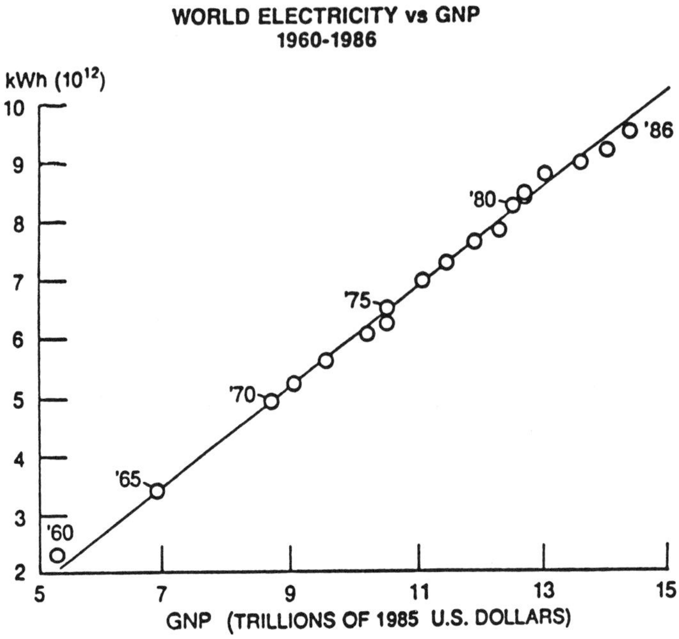

Figure 15

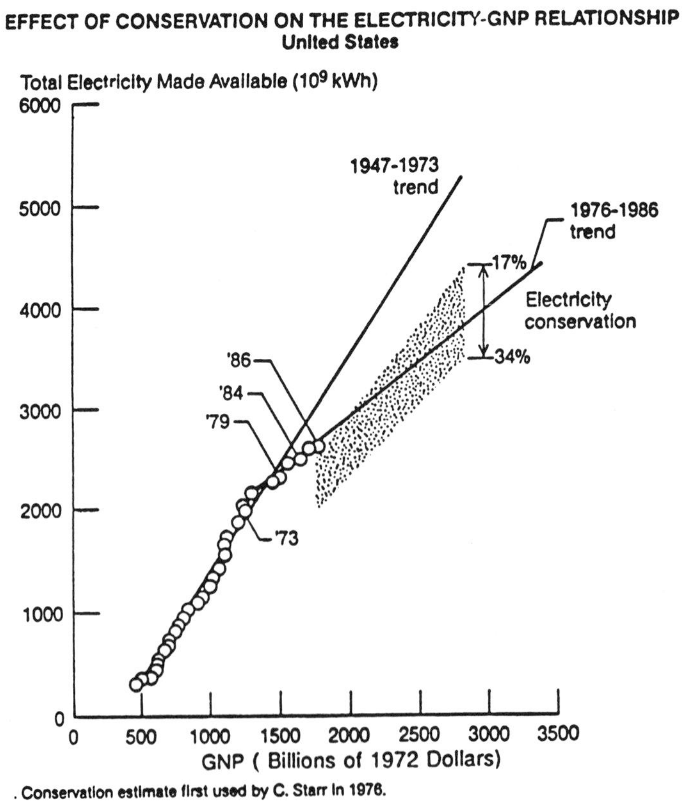

Figure 16

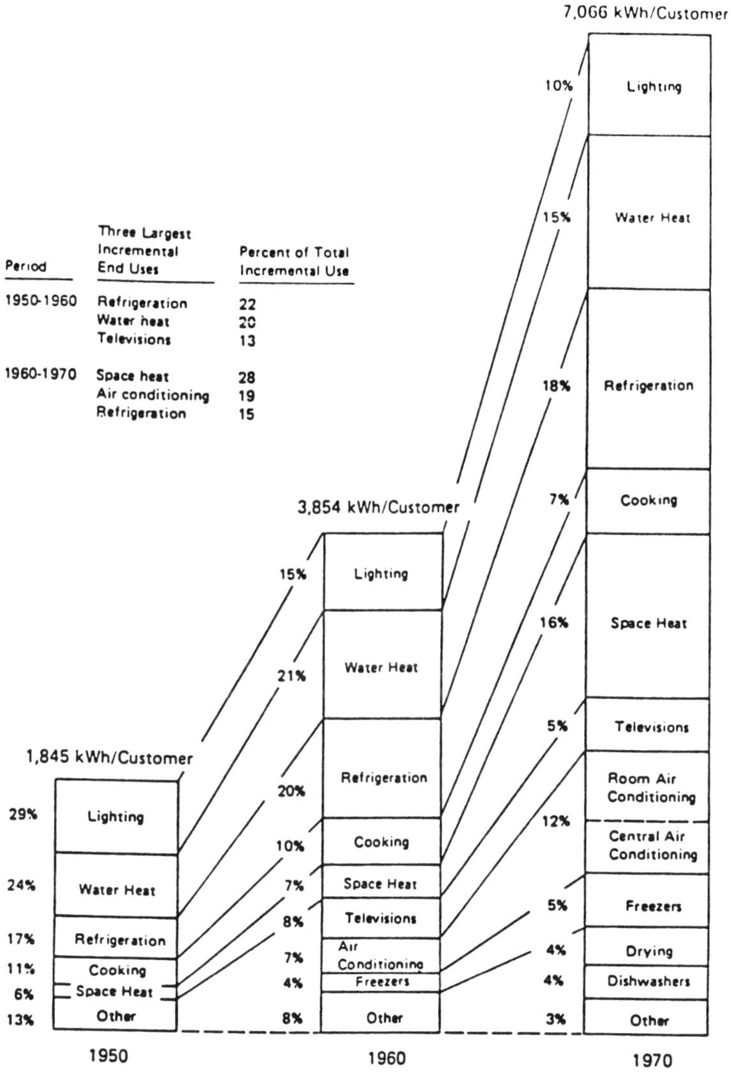

Residential electricity use patterns, 1950, 1960, and 1970.

SOURCE: Tansil and Moyers (1974).

Figure 17

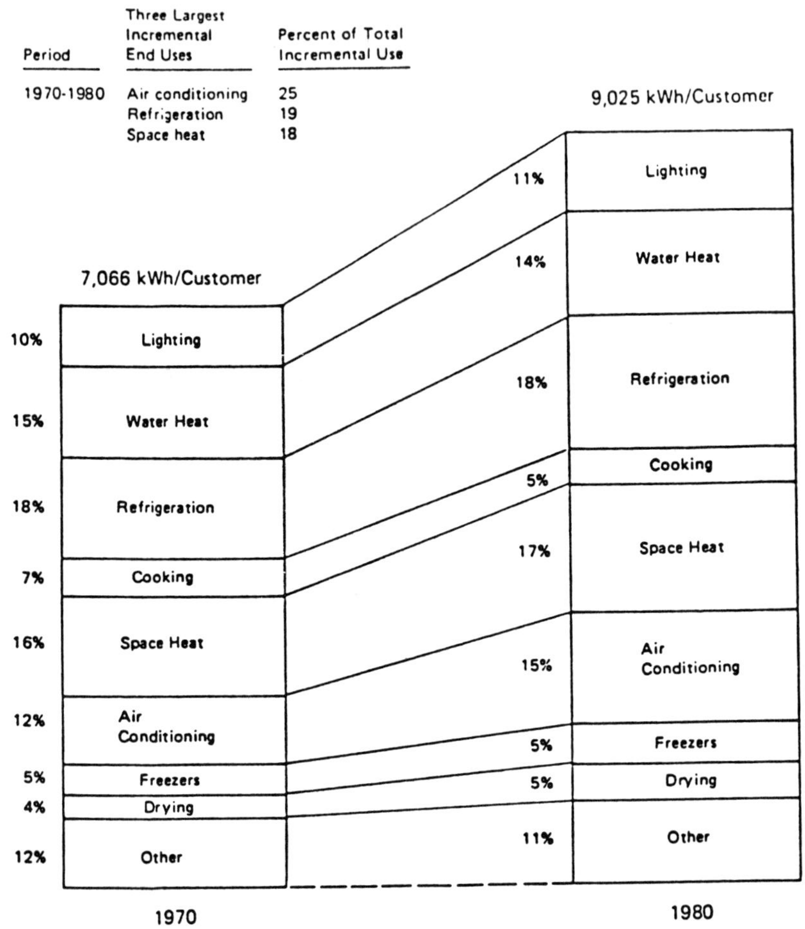

Residential electricity use patterns, 1970 and 1980.

SOURCES: Tansil and Moyers (1974); Geller (1983); U.S. Energy Information Administration (1982).

Figure 18

Very few analysts were as prescient as Chauncey Starr in predicting an increase in electricity demand. There are some that still have not realized its implications.

Fuel Efficiency Improvements: 1973-1988

It is important to understand how these fuel efficiency improvements occurred, so that we may be guided in any projection for the future. I note here an important logical point. We have always been steadily improving fuel utilization and the pre-1973 projections took this into account in a general way. For example, the early electricity power stations put out, as electricity, only 10% of the heat that came from the fuel! This is now 30% or more. Since 1973, as noted, the rate of introduction of efficiency improvements accelerated. The achievements have been great. But we must be careful to avoid "double counting."

It has become conventional to take a pre-1973 energy projection as a base case. If we estimate the fuel efficiency savings that will occur, and subtract them from the base case we may end up by subtracting too much, because the base case already included some of these efficiency improvements. We must be careful not to subtract the improvements that are already subsumed in the projection.

Another standard procedure is to take a projection, probably a pre-1973 projection, and to subtract from it the actual energy usage, or a more reasonable projection, and call the difference "energy conservation savings." (In passing I note that I prefer not to use the expression "energy conservation;" it is too vague. The only precise meaning that has ever been given was by engineers and physicists 150 years ago who proved that "energy is conserved." The discussion here is just another example of its vagueness. Unfortunately the phrase is in common use in today's energy policy discussions.) The procedure makes an enormous assumption of what would have been used without "energy conservation."

I also note that anyone who desires, for political or other considerations, to exaggerate the role of fuel efficiency can do so by choosing initial projections, to compare with the "actual" use, that are unusually high.

Figure 19 shows the historical projection and the actual primary energy use as compiled by the U.S. Department of Energy Conservation. I call to your attention the sudden drop in primary energy use at the end of 1973, when the Yom Kippur war had led to a fourfold rise in oil prices, and another decline at the end of 1979 when the Iraqi invasion of Iran had led to a further rise in oil prices from $12 to $36 a barrel. If we look in more detail (Hogan, 1979), we find that there were probably two distinct steps in this increased energy efficiency. In 1974, industry noticed a large fractional increase in *wholesale* fuel prices. Industry is sensitive to wholesale prices, has the money to pay for increased efficiency, and for consultants who can tell them what to do. Almost immediately industry achieved fuel savings. Homeowners, on the other hand, buy their fuel on a retail basis. They only saw a smaller fractional increase in oil prices. They did not take steps to improve home heating in 1974. However, they began to react to the second oil price rise in 1979.

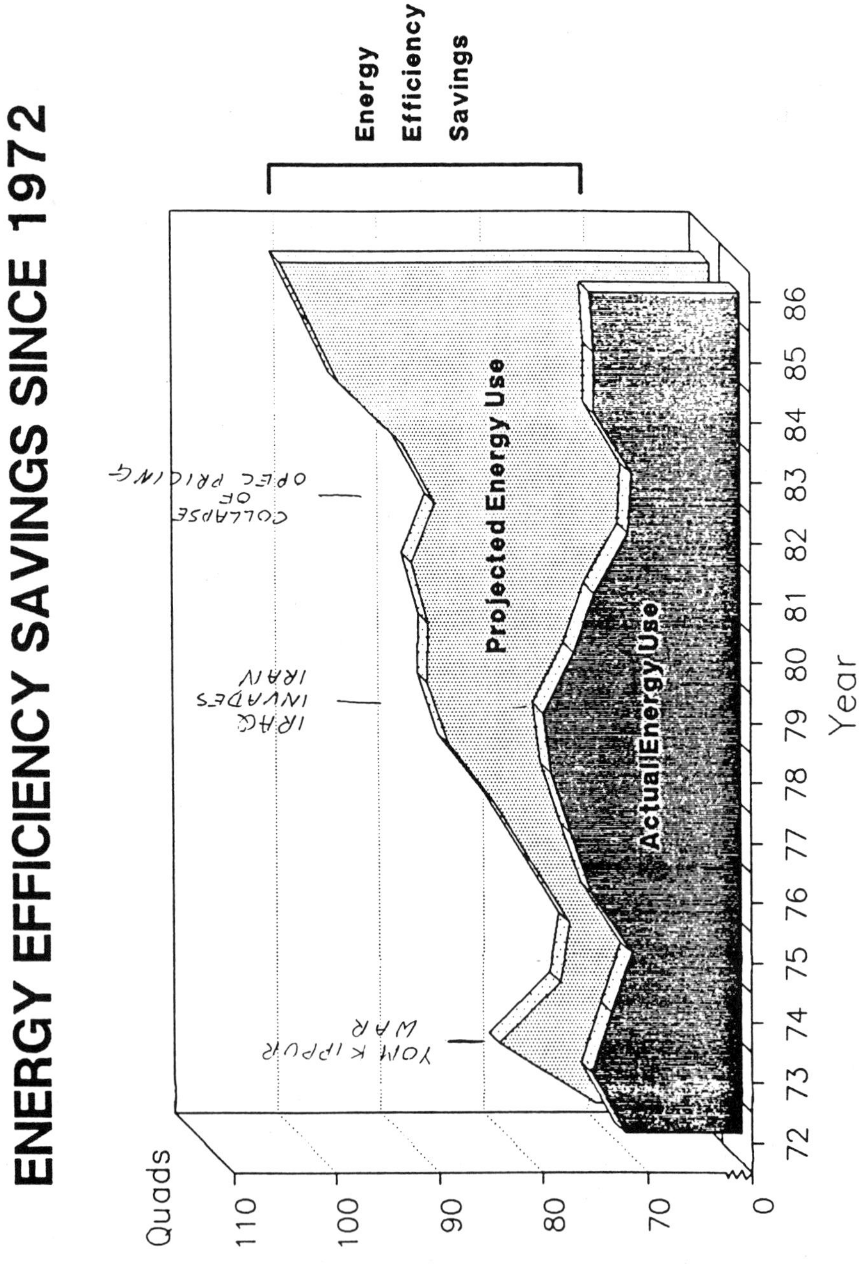

Figure 19

U.S. Department of Energy
Office of Conservation

It was not until 1979 that I met my neighbors buying fiberglass insulation at the hardware store. It was not until 1979 that my hardware store ran out of fiberglass. One conclusion that I draw: it was principally the price increase and the perception of it by the different consumers that led to fuel efficiency.

The above argument does not work for the improvement that has been achieved in consumption of fuels by automobiles. As I will describe in more detail later, the cost of gasoline is only a small part of the cost of running a car, and a part that has been steadily declining – certainly since I arrived in the U.S. in 1955. Perhaps because our elected representatives understood this, automobile fuel standards were mandated by Congress. As we all know, they have been very successful, and have raised the fleet average miles-per-gallon from about 15 mpg in 1973 to 23 mpg today.

We are, however, particularly interested in the possibility of reduction of fossil fuel burning, particularly coal burning. The difficulty of this is particularly apparent from Figure 20, showing that, although the total number of carbon atoms being burned has flattened off in the U.S., it is increasing in China.

I note also two points. If nuclear power did not exist and the electricity was produced by coal, the CO_2 production in 1987 in the U.S. would be 7%–9% higher, as shown. If since 1973 the *increase* in electricity production that was done by coal had instead been by nuclear, the CO_2 production would have been 12% *less*. These nuclear plants could have been built; many were ordered and then cancelled. We could have cancelled coal plants instead, if we had so chosen.

Electricity consumption in the developed countries varies widely. This is shown in Table IV taken from Linden (1987). Norway with its abundant hydro research is high on the list, with Sweden not far behind.

Projections 1989 to 2020

With these thoughts behind us, I now discuss the projections for future energy use in the U.S. and in the world. I first show the projections from the U.S. Department of Energy. These follow the old (pre-1973) method for projections: "take what you have and project" with little thought. The DOE in Figure 21 assumes that total energy use can be held constant, and that the difference from the high energy use can be held constant, and that the difference from the high projection is "energy savings." It is not surprising that they find little change in distribution of energy use either in the U.S. or the world. The U.S. Congress has asked DOE for a plan to reduce CO_2 emissions by 20% by the year 2000, and 50% by 2020. This is the challenge which I will now address.

Potential for Reductions

We see that the major reductions in U.S. fuel use between 1973 and 1985 were in the heating sector. The transportation sector should slightly rise, mostly due to light trucks and vans. There are good reasons for expecting that heating demand *could* continue to reduce and that transportation demand *could* reduce slightly. Transportation in

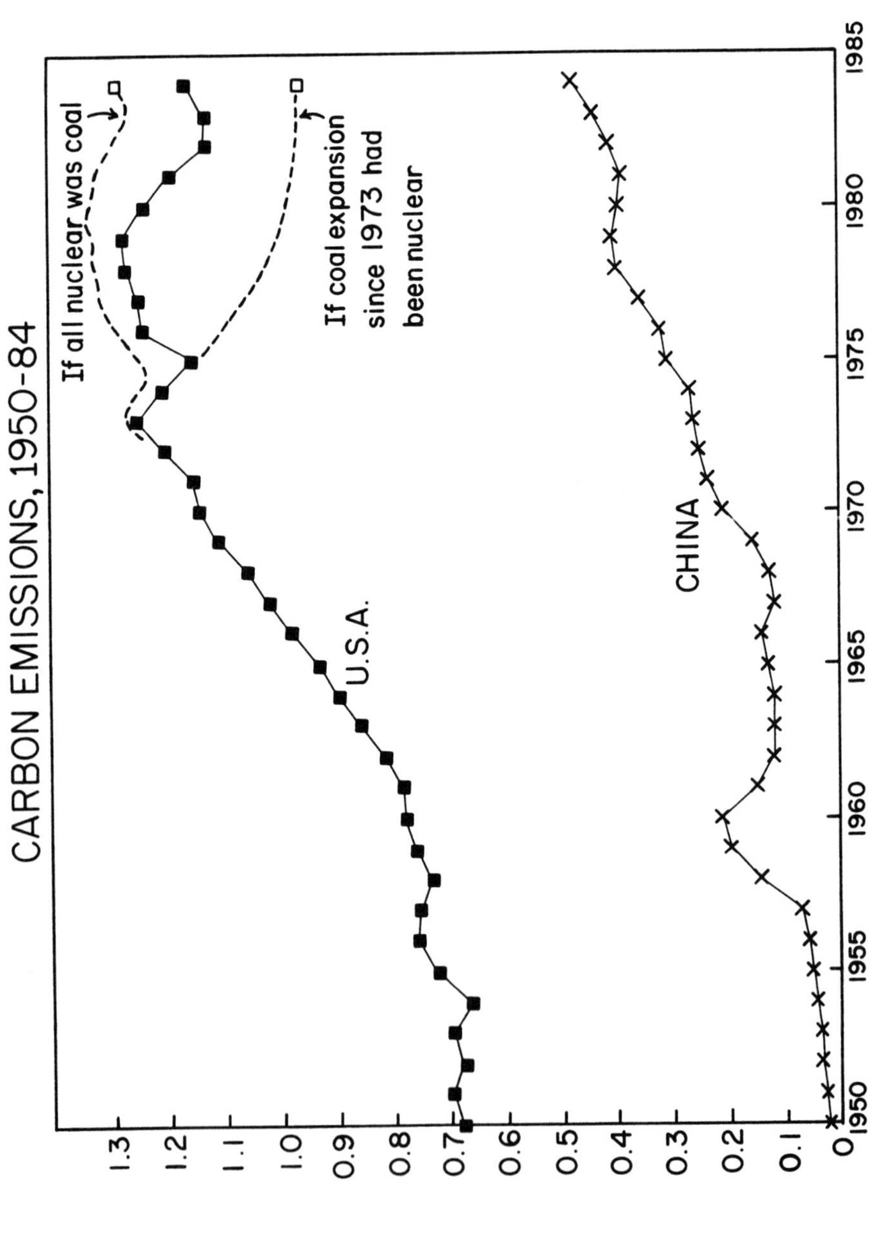

Source: R. Rotty, UN data.

Figure 20

Table IV

Electrification of Energy Supply and
Electricity Intensity of 14 OEDC Countries*

	1985 Share of Electricity in Energy Supply Mix, %	1985 Average Electricity Intensity kWh/ 1972$ of GNP	1950 to 1985 Marginal Electricity Intensity, kWh/ 1972$ of GNP
Austria	19.6	1.5	1.6**
Belgium	15.7	1.2	1.5
Denmark	18.9	1.1	1.5
Finland	29.8	2.5	2.9
France	18.1	1.1	1.2***
Italy	16.3	1.2	1.3
Japan	24.2	1.3	1.4
The Netherlands	12.1	1.1	1.4
Norway	47.6	4.0	4.1
Sweden	41.7	2.5	3.0***
Switzerland	23.0	1.3	1.4
United Kingdom	16.5	1.6	1.3****
United States	17.3	1.5	2.2
West Germany	17.8	1.2	1.5

*The data sources and computation methods employed in this study of electricity intensity of energy supply in OECD member countries over the period 1950 through 1985 and complete graphical and tabular presentations of the results are available from the author.

**1959-1985

***Average of 1950-73 and 1974-85 slopes of kWh versus GNP plots.

****Average of 1950-72 and 1973-85 slopes of kWh versus GNP plots.

FROM: Linden (1988).

the U.S. is largely by private automobile. The fuel efficiency of these automobiles has steadily improved since 1975, from 16 mpg to about 23 mpg. The "half life" of a car is about 10 years and old cars have not all been replaced. Even at present standards more fuel savings will be achieved as old cars are replaced by new ones with more fuel efficiency. There are also new efficient vehicles on the horizon, adiabatic diesel engines and so forth.

However, there are indications since oil prices dropped in 1984 that larger cars are being used and fuel economy is no longer a concern to customers. Perhaps we can persuade Congress to enact more mandatory fuel efficiency standards. But there seems to be resistance to this. It is my view that future fuel savings depend critically on raising the oil price, either directly or by taxation.

The house heating sector offers more future potential for savings. Since 1973 many new products have come on the market: heat pumps, thermal insulation, super E windows, new efficient light bulbs. It has been demonstrated that, even in the northeast United States, solar energy can provide 92% of the energy to heat a house, in addition

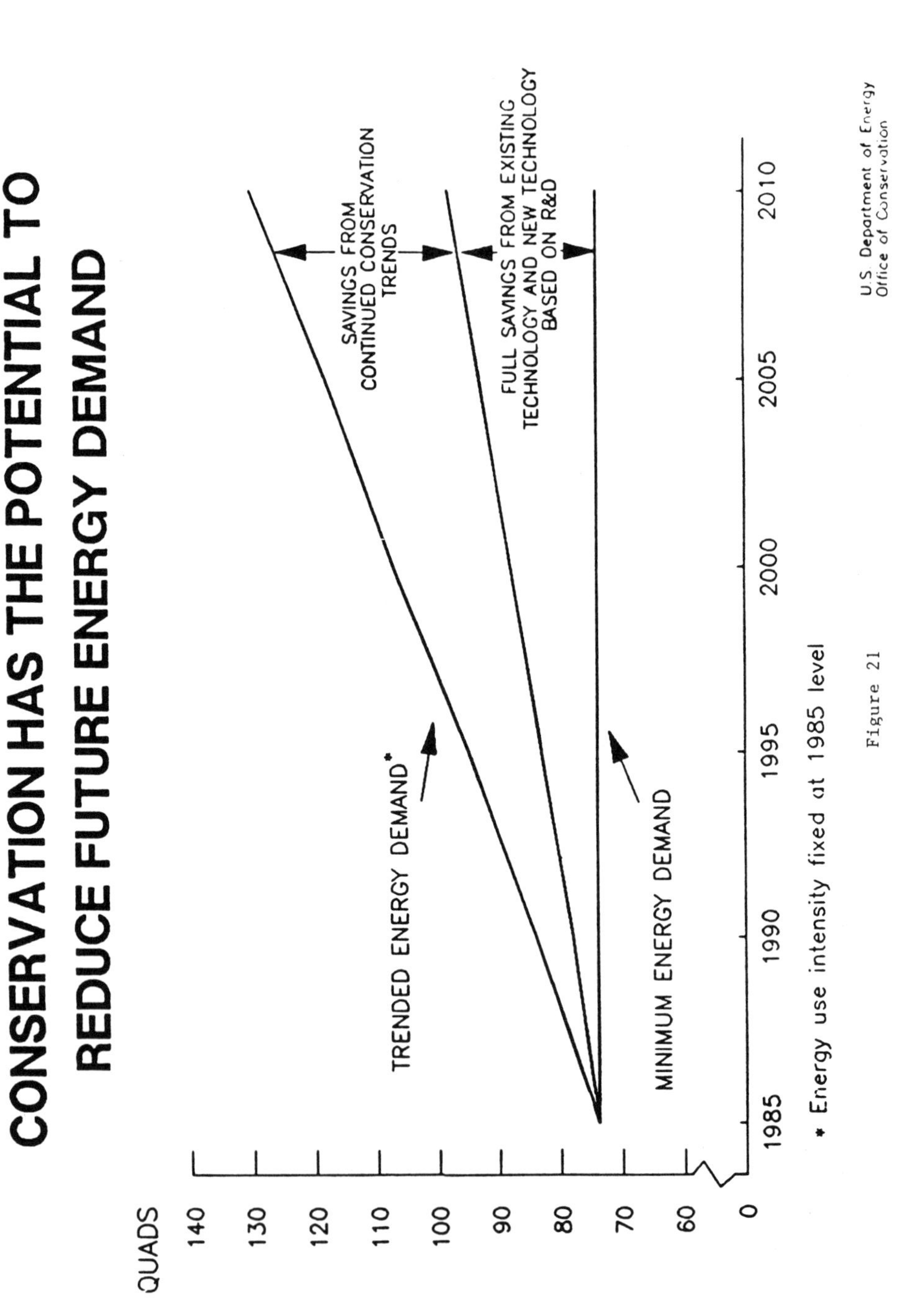
CONSERVATION HAS THE POTENTIAL TO
REDUCE FUTURE ENERGY DEMAND
QUADS
140
130
120
110
100
90
80
70
60
0
1985
1990
1995
2000
2005
2010
TRENDED ENERGY DEMAND*
MINIMUM ENERGY DEMAND
SAVINGS FROM CONTINUED CONSERVATION TRENDS
FULL SAVINGS FROM EXISTING TECHNOLOGY AND NEW TECHNOLOGY BASED ON R&D
* Energy use intensity fixed at 1985 level
U.S. Department of Energy
Office of Conservation

Figure 21

to the heat that comes from household appliances: electric stove, refrigerator, washing machine and so forth (Hyman, 1977; 1978). But this house was capital intensive, and 70% might be a more practical goal. Although there have been savings in house heating already, the lifetime of housing stock is 40 years or so, so more savings can come when housing turns over. I show in Table V a list of space heating requirements for single-family houses in the U.S. and Sweden. I take as a goal that we should, could and will replace all old houses by new ones that use 1/3 of the energy input than was formerly used. This will result in a reduction in home heating requirements.

Table V

Space heating requirements in single family dwellings
(kJ per m^2 per degree day

United States	
Average housing stock	160
New (1980) construction	100
Mean measured value for 97 houses in Minnesota's Energy Efficient Housing Program	51
Mean measured value for 9 houses built in Eugene, Oregon	48
Calculated value for a Northern Energy Home in the N.Y.C. area	15
Sweden	
Average housing stock	135
Homes built to conform to the 1975 Swedish building code	65
Mean measured value for 39 houses in Skane, Sweden	36
Measured value, houses of Mats Wolgast, in Sweden	18
Calculated value for alternative versonsof the prefabricated house sold by Faluhus	
Version 1	83
Version 2	17

FROM: Goldemberg, 1985.

We can therefore hope that in the U.S.A. primary energy use will remain constant or even fall slightly, but only if we take active steps to ensure this. Most analysts do not expect this to occur, or expect it to occur only by a stagnation in the increase of GNP.

However, in the electricity sector there is comparatively little room for increased energy efficiency. Most analysts expect that the fraction of primary energy use that is used for electricity will continue to increase in all countries. This is a fundamental point that leads to the conclusion that I make later that some crucial policy decisions can be separated.

Comparison of Projections

I now compare in Figures 22, 23 and 24 actual world energy use in 1980 with projections for the year 2020 from the World Energy Congress (WEC), International Institute for Applied Systems Analysis (IIASA), and Goldemberg, et al. The first two of these had several projections, differing primarily in the projected GNP growth. Although both WEC and IIASA carefully analysed the way in which energy is used, this was primarily reflected in the ratio of energy to GNP discussed earlier; in each of their projections they expect energy/GNP to fall by about the same amount, but if there is low GNP growth, we will not go so far along the projected curve before 2020.

If we take the mean of the low and high projections of WEC and IIASA, they still project twice the primary energy use in the industrial countries, much more in the primary energy use in developing countries. Goldemberg, et al.'s study is primarily a study in end use. They emphasize that they are making this not as a projection of what is likely to happen based on historical trends, *but* as a *strategy* for society to organize itself if it wishes to change this historical trend. I take Goldemberg, et al.'s study as representative of a number of "low energy growth" studies. It is not polemical, and seems to be reasonably argued. Most of the other low energy growth studies do not satisfy these criteria; one projects a *reduction* of fuel use in the developing countries per capita use of energy below the present 400 watts! The lack of moral and other sense in such a projection makes it hard to accept. In Tables VI and VII I show more details of the Goldemberg, et al. projection.

I now note one feature that is not emphasized by Goldemberg, et al. They project a global doubling of electricity demand. This is in accordance with my arguments earlier. Yet they do not properly discuss the options on how this demand can be met. In Goldemberg's study, the breakdown between nuclear and fossil fuels in electricity generation is almost arbitrary. They attempt to keep the nuclear contribution in their scenario small. This they justify by their desire to reduce the incentive to make a breeder reactor producing plutonium, or using plutonium recycle. They have been concerned that separating plutonium in the world energy economy can lead to terrorism or war. This may be a correct prediction of peoples' concerns, and society's reaction to them, but they fail to emphasize that this is a separate decision and should be considered in its own right.

It must be noted that by 1985 most of the coal used in the U.S. was in electricity generation. Coal is used only in large base-load plants which are capital intensive and run all the time. Natural gas is scarcer and it is used primarily in peaking plants, used only for a portion of each day. Natural gas is also gaining acceptance for use in base-load plants because natural gas plants can be built rapidly to correct for lack of foresight in installing adequate base-load capacity. Most of these coal-fired stations will be replaced by 2020, and a replacement by nuclear plants is technically possible. If we accept

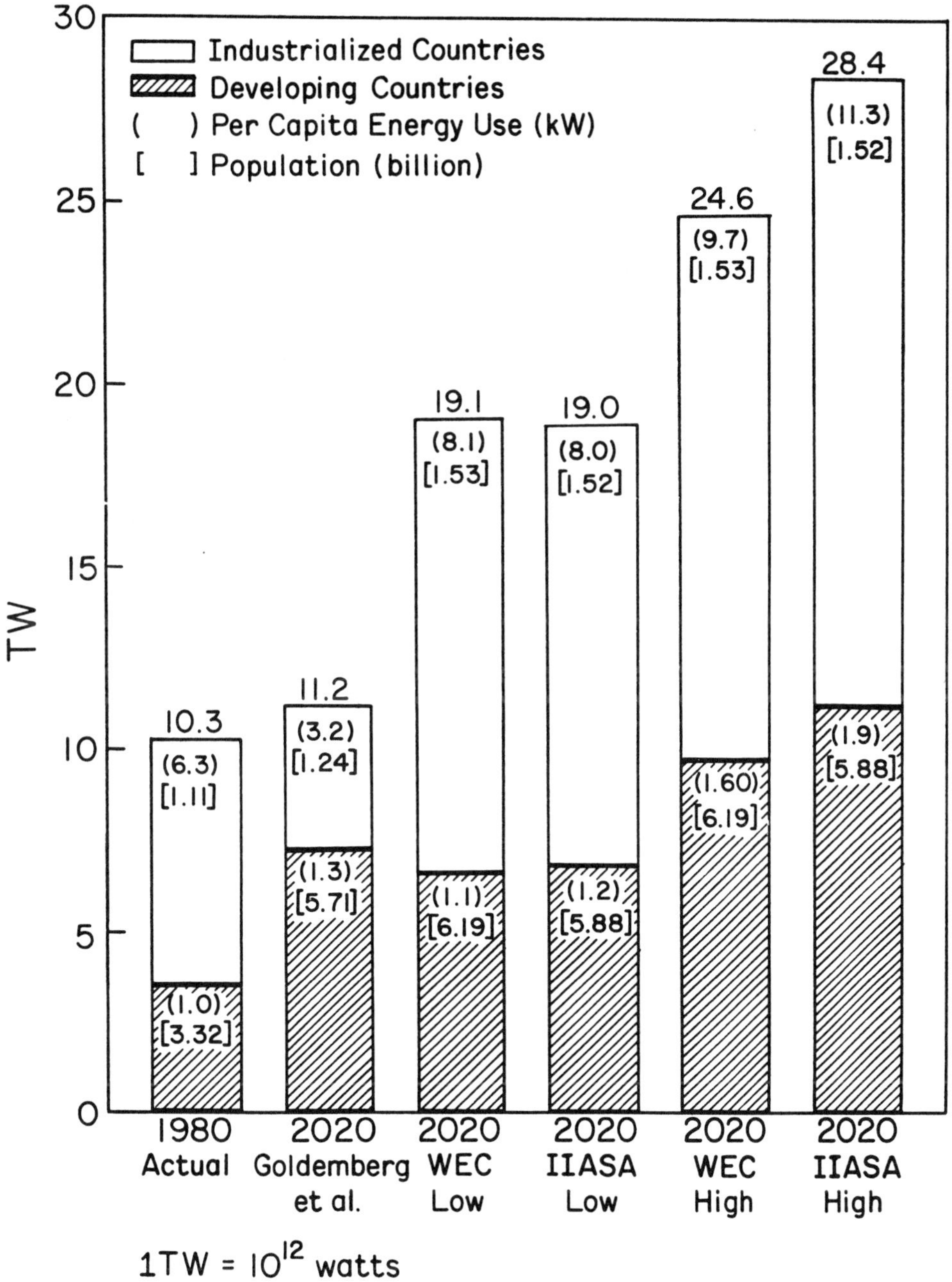
Industrialized Countries
Developing Countries
( ) Per Capita Energy Use (kW)
[ ] Population (billion)
TW
30
25
20
15
10
5
0
10.3
(6.3)
[1.11]
(1.0)
[3.32]
11.2
(3.2)
[1.24]
(1.3)
[5.71]
19.1
(8.1)
[1.53]
(1.1)
[6.19]
19.0
(8.0)
[1.52]
(1.2)
[5.88]
24.6
(9.7)
[1.53]
(1.60)
[6.19]
28.4
(11.3)
[1.52]
(1.9)
[5.88]
1980 Actual
2020 Goldemberg et al.
2020 WEC Low
2020 IIASA Low
2020 WEC High
2020 IIASA High
1TW = 10^{12} watts

Figure 22

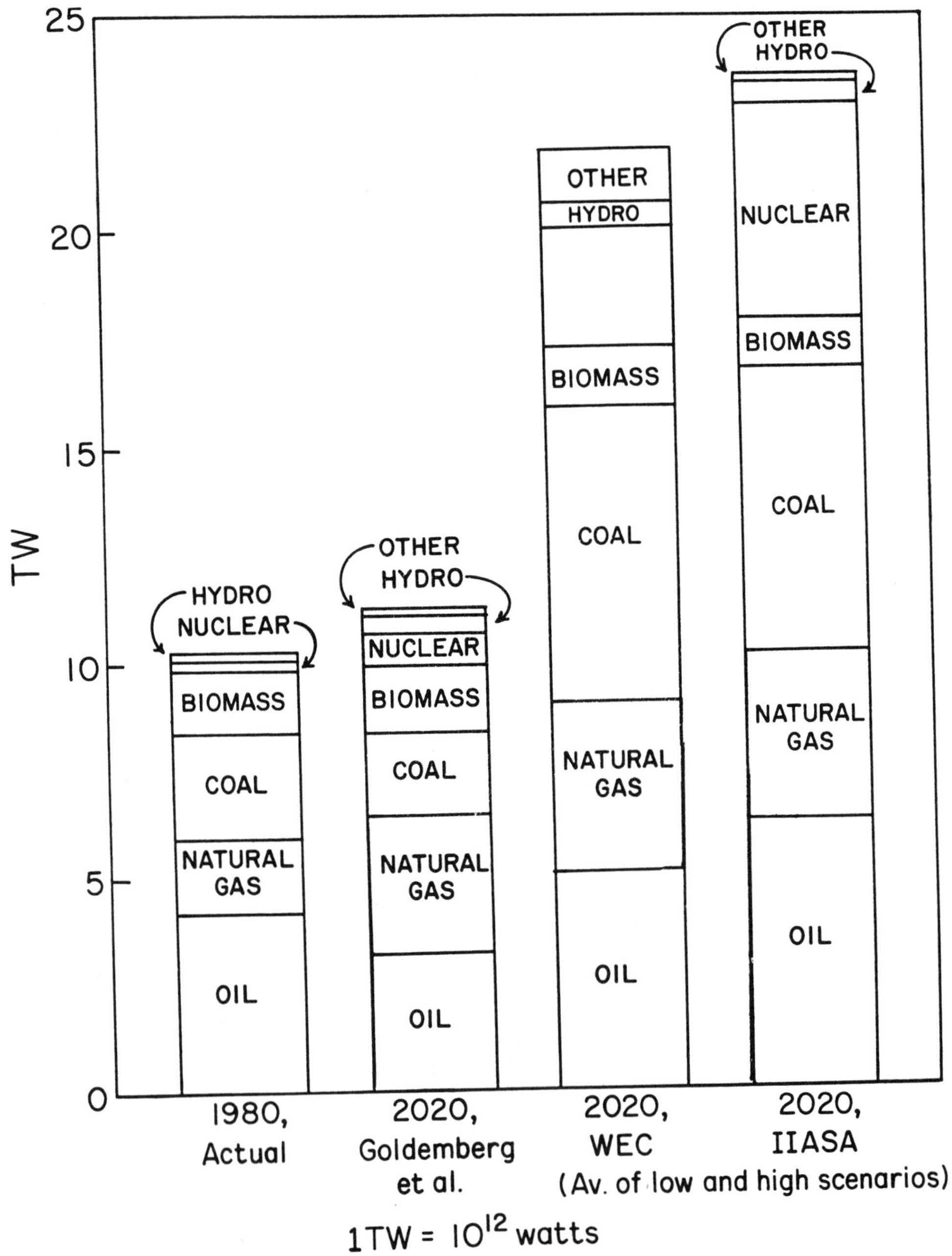
25
20
15
10
5
0
TW
HYDRO
NUCLEAR
BIOMASS
COAL
NATURAL GAS
OIL
OTHER
HYDRO
NUCLEAR
BIOMASS
COAL
NATURAL GAS
OIL
OTHER
HYDRO
BIOMASS
COAL
NATURAL GAS
OIL
OTHER
HYDRO
NUCLEAR
BIOMASS
COAL
NATURAL GAS
OIL
1980, Actual
2020, Goldemberg et al.
2020, WEC
2020, IIASA
(Av. of low and high scenarios)
1TW = 10[12] watts

Figure 23

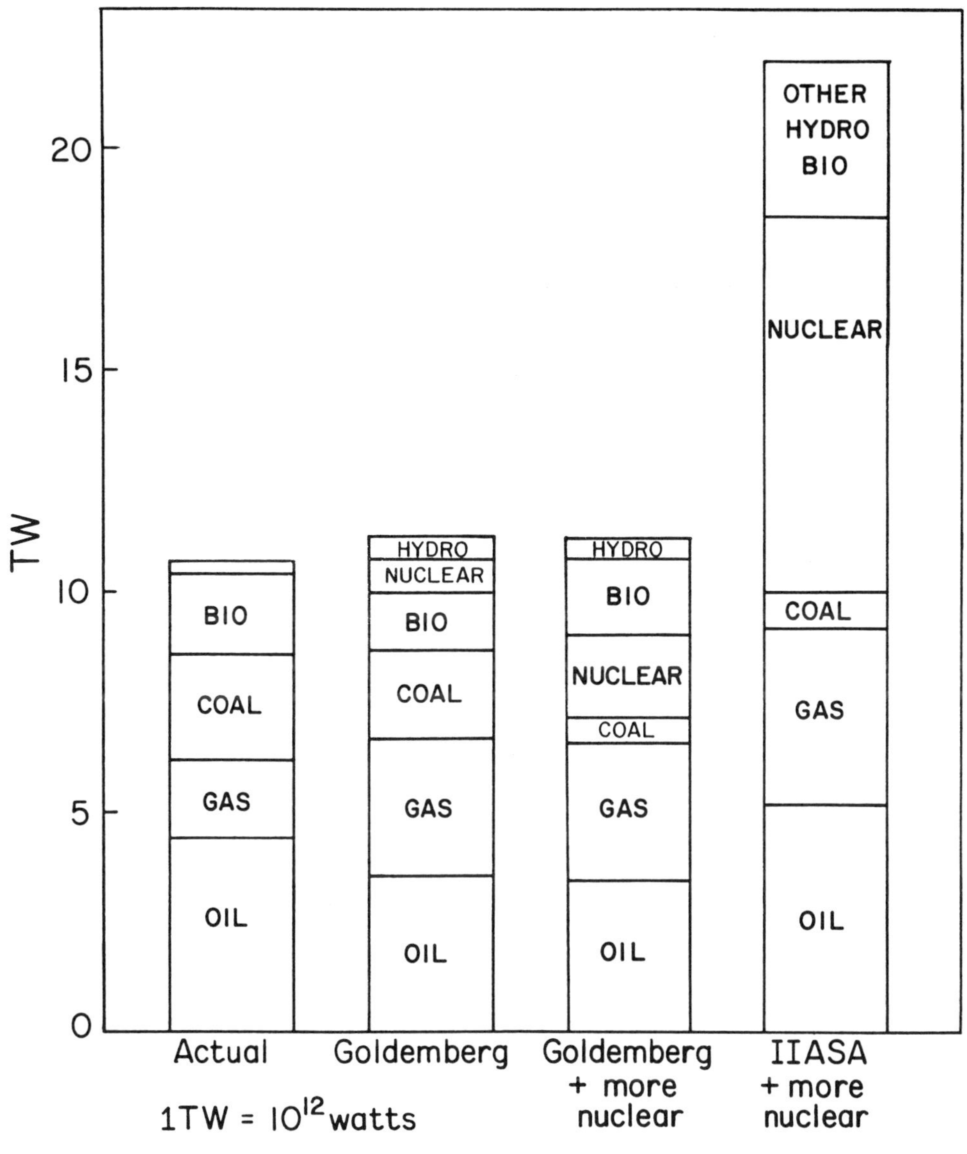
TW
0
5
10
15
20
OTHER
HYDRO
BIO
NUCLEAR
COAL
GAS
OIL
HYDRO
NUCLEAR
BIO
COAL
GAS
OIL
HYDRO
BIO
NUCLEAR
COAL
GAS
OIL
BIO
COAL
GAS
OIL
Actual
Goldemberg
Goldemberg + more nuclear
IIASA + more nuclear
1TW = 10^{12} watts

Figure 24

Table VI

Global primary energy supply scenario (in TW)

	1980	2020
Nuclear power	0.22	0.75
Hydropower	0.19	0.46
Wind and photovoltaic electricity	----	0.09
Fossil fuels		
Coal	2.44	1.95
Oil	4.18	3.23
Natural gas	1.74	3.23
Subtotal	8.36	8.41
Biomass		
Organic wastes		0.74
Plantations or wood farms	____	0.75
Subtotal	1.49	1.49
Totals	10.3	11.2

FROM: Goldemberg, 1985.

Table VII

Global electricity supply scenario (in TW)

	1980	2020
Hydropower	0.19	0.46
Wind and photovoltaic electricity	----	0.09
Cogeneration		
Biomass	----	0.14
Fossil fuel,	----	0.13
Central station		
Nuclear	0.08	0.30
Fossil fuel	0.66	0.66
Totals	0.93	1.78

FROM: Goldemberg, 1985

NOTE: This is not the primary energy.

Goldemberg's scenario but replace coal burning by nuclear power, the fossil energy used is reduced by 2TW, 25% of the fossil fuel energy use, leading to a 30% reduction in carbon emissions. There have been claims and counterclaims that, on the one hand, nuclear energy can reduce CO_2 emissions and, on the other hand, fuel efficiency can do better. It is important to realize that the decision on whether to proceed with nuclear energy or whether to encourage fuel efficiency are, in all these scenarios, almost independent questions.

Although by 1985 there was little oil used in electricity generation in the U.S., in some regions about half the electricity is from oil. This applied to the coastal northeast and to California. Whether this is because imported oil is especially easy to obtain on the coast or whether it is because the residents of California and Massachusetts especially object to coal and nuclear power is unclear.

The fact that coal and nuclear energy are almost interchangeable in the present and foreseeable future U.S. energy mix has been used, before the greenhouse effect came to public attention, as an argument that nuclear energy is unnecessary and that, instead, we can use "clean coal." It is hard to remove more than 90% of the sulfur, harder to remove nitrogen oxides. And no amount of cleaning coal can get rid of carbon dioxide.

That society should reject nuclear power and use clean coal was the claim, remarkable for an environmental group, of the Sierra Club's "energy policy" promulgated in 1981. Such claims led Paul Tsongas, then Senator from Massachusetts to respond that he did not see how anyone who prefers coal to nuclear energy can be correctly described as an environmentalist.

Some authors try to ridicule the idea of replacing coal-fired power plants by nuclear plants by describing what has to be done. If all central station power in Goldemberg, et al.'s scenario is to be replaced by nuclear power, 0.96 TW of generation must be built or about 1,200 nuclear plants; about 400 would be in the U.S.A. The U.S. built nearly 100 plants in the last 15 years; another 50 or so were ordered and cancelled and we could probably have built 150 or 200 if we had so desired. Another 200 in the next 30 years is easy if the political situation permits it.

I note that I leave out some possible options; it is possible to switch some more energy end uses from direct fossil fuel use to electricity and thereby enable uranium to supply the need. This can be done by more electrical heating of houses or by increasing electrification of transport. Fifty years ago in England they were the usual milk delivery vehicles. Electric cars are possible, and have been used for local transport. Railroads can electrify, although the conversion of the Milwaukee Railroad's mountain line from electric to diesel in 1974, followed soon thereafter by abandonment, encourages me to leave this out as a likely option.

Means to the End

If we set ourselves the goal of *reducing* global CO_2 emissions, it seems that we must not only work as hard as we can on efficient energy utilization as Goldemberg et al. suggest, but also on conversion to nuclear energy. I noted that fuel efficiency in the past has been encouraged by high prices and I have stated my belief that that is the main reason for increased fuel efficiency since 1973. This suggests a tax on fuels, and particularly on those fuels whose use we need to restrict—coal, oil and gas. A uniform tax per carbon atom seems appropriate. Such taxes have been discussed by Nordhaus (1980) and Nordhaus and Yoke (1982) who calculate the effects of a tax of $10 per ton of coal (in 1975 dollars). This is a small tax. A more reasonable tax of $10 per ton of coal (in 1988 dollars) might be: $0.25/gallon of oil; $70/ton of coal; $0.1/100 cubic feet of gas. Most analyses of efficient fuel utilization find that there is no one panacea; decisions have to be made by thousands and millions of individuals. A free market economy guided by taxes is likely to work better than an authoritarian one with compulsion. The high value of the energy to GNP ratio in the U.S.S.R. that was noted earlier (Figures 10 and 11) may be an indication of the inability of a central government to make the large number of different decisions. A carbon tax serves a different purpose than the tax on oil or only imported oil, proposed by some, that is intended to encourage domestic supply of oil.

But I now illustrate the limitations that even price can give an incentive by comparing in Table VIII costs in 1950 and 1985. Relative to a typical salary of which I am cognizant, the gasoline price is halved and, weighted by a car mileage, it has gone down nearly threefold. Gasoline price was not much of an incentive (by itself) for fuel economy in 1950. Why should it be in 1988?

Table VIII

Gasoline Costs Relative to Other Costs

	1950	1988
Assistant Professor's Salary (yearly)	$ 5,000	$30,000
Cost of car (new)	$ 1,000	$10,000
Gasoline price (per gallon)	$ 0.25	$ 0.90
Efficiency of a car (mpg)	15	25
"Naive incentive" (price/salary) ÷ 100,000	5	3
"Sophisticated incentive" (price/salary) ÷ mpg	1/3	1/8
Typical cost of gasoline for 10,000 miles	$ 167	$ 360
Typical yearly amortization, insurance and repair costs	$ 500	$ 2,500

In particular, the last row shows that *if* fuel efficiency adds to unreliability or maintenance costs generally, the incentive will be in the opposite direction. Fuel would (and probably should) be used to reduce other costs.

To operate in a free society, consumers need good information. One of the most important parts of the DOE programs, and one in which the Center for Environmental Information can help and has helped, is to ensure that reliable information on fuel saving procedures is widely disseminated. High prices and information may not be enough. Sometimes efficient energy utilization is discouraged by institutional barriers. Institutions that assign some of the department activities to an overhead pool immediately remove the direct incentive to make fuel efficient. Gas, electricity, heat and light are traditionally provided by the institution to everyone. They are in the "overhead" base. If we get an institutional change in universities, schools and government buildings, we will achieve almost immediate fuel savings. An example of the effect of removing such a barrier can be found at my own university. The Division of Applied Sciences has a special arrangement that they "keep their own overhead" and heat their own buildings. Any fuel efficiency improvements are immediately seen by the faculty as more money to spend. In the years 1974-76, Professor Abernathy showed how to reduce fuel input to the building fourfold with little change in comfort. Professors in other parts of the university have far less incentive.

Strategy Projections for Energy Use in the Next 100 Years

In this section I consider what strategy we can have beyond 2020. These are more problematical and uncertain than the discussions so far. I here note some general features only and go into detail on a few uses.

Population Reduction

The most important uncertainty, and the largest potential for restricting world energy demand in the long term, is to control the world population. In the developed world population growth is (almost) under control and only a small increase is anticipated by 2020. However, large increases are still occurring in the undeveloped world. Although the per capita energy use is small, this is multiplied by the large population. By 2020 both per capita energy use and the population might increase enough to make developing countries use half the global primary energy. Therefore, the most effective long-term reduction in fuel use might come from birth control. In this, developed nations could well offer as much help as the developing nations desire. However, the options for controlling birth raise as much discussion and as many disagreements as the nuclear/other energy discussions.

Nuclear Power

To reduce the use of fossil fuels further, some would like to see a definite policy like that in France, that the U.S. will never again build a central station electricity generating plant using fossil fuels. The 1987 repeal of the law restricting the use of natural gas in central station electric power noved the U.S. in the opposite direction. What are the needed requirements to have an increase in nuclear power? This was so well described by Uhrig (1986), formerly Vice President of Florida Power and Light Co. and now Professor at the University of Tennessee, a utility company executive who believes that nuclear power is cheaper, environmentally superior to other alternatives, and in the public interest. He will nevertheless not order a plant at the present time. He will not order one until four conditions are met:

1. "When I know what the plant is going to look like (and there will not be hundreds of NRC ordered retrofits)."
2. "When I know what it will cost."
3. "When I know when it is going to be finished."
4. "When I know that I can operate it after it is finished."

and I add a fifth: "When I know that I can recover the construction cost during operation." These do not seem to be unreasonable conditions.

He knows none of these things now. In each of these features, nuclear power is treated more strictly than other fuels.

Nuclear power is presently restricted by a combination of:

1. Vocal public opinion;
2. Regulation;
3. Political consideration.

Electricity from nuclear power has, in the past in the U.S. and in France today, been cheaper than electricity from coal. The added regulations and other constraints have added to the price. Some will say that these constraints are necessary for environmental protection and for public health; if they cannot be met, they will argue that we must discard nuclear power as an option and accept the 10-30% increase in CO_2 emissions. Others will say that the constraints on nuclear power are so much more strict for coal, that they lead to a net reduction in environmental protection and health even now.

Perhaps the nuclear power industry can redesign the parts of the system that cause people concern. Then the consequent change firstly in public opinion, then in regulations, will allow an expansion. This might involve:

1. Reactors with more inherent safety;
2. Remote siting;
3. Demonstrated waste disposal.

There are many indications that a resurgence of nuclear power may need federal control. The utility industry is contemplating a restructuring which would separate the wholesale generation of electricity and the retail sales of electricity to the ultimate consumer. This may be an occasion to modify the position taken in the Atomic Energy Act, where private companies engaged in nuclear powerplant construction. This could be undertaken by the government as in most other countries. These are the challenges for U.S. citizens, our new administration, and for the whole world.

Dr. Andrei Sakharov, whose leadership in matters of civil rights and human needs is unquestioned, repeatedly goes out of his way to emphasize the importance of nuclear power not only for the developed world, but also for the developing world (Sakharov, 1978; 1987; 1988). He has called upon antinuclear activists to spend their energies instead on making nuclear energy safer.

Natural Gas

In 1973 the U.S. seemed to be running out of natural gas. Most of the gas had been discovered along with oil. Most of the oil in the U.S. had already been discovered, and it was thought that the gas had been discovered within it. In the intervening 15 years, much more gas has been discovered, mostly in places where there was no oil. This has suggested to many that natural gas might make a major contribution to future energy supplies. In particular it seems that the supplies in the U.S.S.R. can provide 20% of Europe's energy needs for the next 50 years. However, the detailed projections for natural gas have not matured sufficiently to be sure of them. Burning natural gas produces half the carbon dioxide per unit of energy than coal produces.

It had been thought that coal, oil and gas were fossil fuels. However, Thomas Gold has suggested that most natural gas has a primordial origin, and there is 100 times as much in the center of the earth as had previously been discovered. Although it is unclear whether or not this is true, it is now clear that this gas, if it exists, is not easily discovered.

It appears, therefore, that we might not be able to count on natural gas to replace coal and oil in Europe's fuel supply. It seems less likely in the U.S. However, this is only half a solution to the CO_2 problem as compared to using nuclear powered electricity. Moreover, natural gas will burn well in small burners in households and small businesses with no more environmental damage than its use in big burners. If there is a conflict, natural gas should be reserved for domestic heating and cooking, where it gives the maximum environmental advantage.

Solar Energy

It may be noted that I have not explicitly mentioned many uses of solar energy in these projections. There are many reasons for this. Firstly, wise use of the sun is, and always has been, implicit in all energy planning. The subject of this conference is changes in the climate as the delicate balance of the heat of the sun that impinges on the earth and infrared heat that leaves the earth is upset. It is obviously a sensible principle to upset this balance as little as possible and only after considerable thought. To interrupt the sun's heat, on its way to warm the earth, use it to man's purposes, and then to release it to do its final task of keeping the earth warm, seems an excellent procedure. When this is done, it is described as using solar energy. I usually count 20 feasible ways of using the sun, some of which are economic in some places, some in others, and some not economic at all.

From Figure 2 it can be seen that before 1880, the U.S. was primarily a "solar economy" depending upon wood and passive solar heating. The U.S. went away from this solar economy, not because we were deliberately trying to be wasteful, or deliberately polluting, but because coal and oil were cheaper and more convenient. Even a cursory examination of costs suggests that coal and oil have become *cheaper* even in the last 40 years, their use in society more efficient, and less polluting. This suggests a simple test for anyone who wants to return to a solar technology: what improvements have been made in this technology since 1850? If there have been few improvements, it will not compete in the marketplace today.

The use of wood (and other biomass) is an example. A large number of people in the affluent suburbs bought wood stoves in the 1970's, but I have seen that many have now been thrown out as there was overoptimism about the cost of burning wood. But in countries other than the U.S., wood can and will be important. In all the energy projections for the world that I showed in Figures 22, 23 and 24, biomass (including wood) was considered to play a major role, but primarily in developing countries.

Figure 2 showed the importance of hydroelectric power. This is a form of solar energy; it is the sun which raises the water as vapor, to descend as rain. It is collected in dams and falls through pipes to water turbines to generate electricity. In developed countries most hydro sources are developed already. However, all the projections plan for hydropower increases in developing countries.

As we discussed earlier, some people consider fuel efficiency improvements as "energy sources" and on the left-hand side of a supply-demand equation. More conventional historically, and in this paper, I consider them to be a reduction in demand. So it is with passive solar heating; the more we heat our houses with the sun, the less demand for other fuels. This has always been the case and passive solar heating is not new. I therefore have not considered passive solar energy as a source of energy for these projections. But as a major component of a reduction in demand, passive solar heating of buildings is already included in the fuel efficiency discussion.

It is generally agreed that other solar technologies, such as windpower and photoelectric production of electricity by solar cells, are not likely to contribute more than a percent to the global fuel supply before 2020 and only appear as a thin line in these projections. However, after 2020 they may become important. In this I note a conclusion in a report of the American Physical Society that with silicon cells the mounting structures are likely to be extensive (because of the low energy density) and cost more than the cells. This suggests a detailed *systems approach* to consider places where solar cells are most likely to make a contribution.

As I noted in the introduction, windmills have been used for 1,000 years. A century ago, windmills were used in almost every farm in the west and midwest. Between 1900 and 1970, many more windmills were taken out of service than were put into service as gasoline engines were cheaper. In 1973, it seemed to me, among others, that the age of the windmill was over. I was wrong. Figure 25 illustrates how another person, as described by Cervantes, was confused about windmills. New aerodynamic construction makes windmills more efficient than before.

Windmills, in certain locations, have begun to make an impact on electricity generation. In large part this was done by massive subsidies and tax rebates. It is a standard joke that anyone who put a windmill in Altamont Pass in California could make money whether it worked or not! A more realistic view is that the subject is tangled enough that it is hard to tell whether anyone would make money without subsidies. Now subsidies are reduced or withdrawn; it will take another 20 years or so for a realistic economic assessment to be made. I am optimistic that in some locations, such as Altamont Pass, Hawaii, the Shetland Islands and the Caribbean, where the wind can be strong and steady, windpower will be economic and by 2020 make a major local contribution. Worldwide it will not make a major contribution before 2020 and I personally doubt whether it will ever do so.

Recently it has become an article of faith that these technologies are especially suited for developing countries (Schumacher, 1973). However, this is far from clear. For example, in Egypt attempts were made in the 1950's to introduce windmills. The Mediterranean coast seemed ideal for their use; the wind is strong and continuous. However, the project collapsed on the issue of maintenance. Thousands of windmills need many technicians for their maintenance. A developing country may not be able to find the many technicians needed.

By contrast, a centralized power station (such as a nuclear power plant) needs comparatively few qualified technicians and engineers, which a developing country can often acquire. This is an important subject for debate which includes facets that go beyond the subject of this paper. For example, can a developing society withstand the separation into the trained and untrained that the last sentence suggests?

I specifically *reject* some suggestions. In particular I believe that much discussion in the U.S. of alcohol fuels is wrong. It is mainly motivated by questions of monetary gain for a few. In Brazil, making ethanol (alcohol) from sugar cane makes some sense since they have the cane that otherwise would go to waste. Making ethanol from corn in the U.S. makes no sense. We spend more fuel to make the fertilizer for growing the corn than the energy content of the ethanol! We also produce N_2O which goes to the upper atmosphere and causes havoc in this unnecessary process. I regret that both our new President and his opponent supported ethanol use in one of the debates. But the tax credits that have been given by thoughtless federal and state governments have allowed entrepreneurs to make money in spite of the net energy loss.

I agree, however, that environmental control in a city may be easier if automobiles burn alcohol. Alcohol burns more cleanly with less nitrogen oxides and there are therefore good environmental advantages. There is good limited reason to support alcohol fuels.

"Synfuels," and in particular oil made from coal, were touted ten years ago as a replacement for oil. This was usually considered to be a procedure for making synthetic oil – as was done in several countries in World War II when oil supplies were temporarily interrupted. Linden (1987) and others have argued that to make oil from coal as a long-term policy is unnecessary, expensive, and stupid. It makes more sense to use *methanol* (not ethanol). It is easier to make methanol from coal than oil or gasoline and methanol has the same environmental advantages in burning that ethanol does. Methanol therefore makes some sense for a long-term energy strategy to replace oil. But this is interesting and important for other environmental reasons and for reasons of fuel supply.

The conversion of coal to methane and ultimate burning will, of course, *increase* CO_2 concentrations.

Conclusions

There are many pressures that can cause world energy use, including fossil fuel use, to expand considerably over the next 20 years and the next century. The increase in demand in developed countries has been reduced, but in developing countries both the population and fuel use per capita are likely to increase. It will probably take explicit steps to resist these pressures and reduce fossil fuel burning.

There is still considerable scope for increased fuel efficiency. The primary incentives for fuel efficiency have been price and cost increases. This raises the issue of how to achieve these incentives. These improvements are most likely in the non-electricity sectors.

The trend to electricity use continues with, and is partially caused by, a reduction in overall energy use. In this sector it is easy to replace coal- or oil-fired power plants by nuclear generating plants, if we wish, and thereby reduce CO_2 production.

Acknowledgments

I am particularly grateful for helpful discussions with C. Starr, J. McGranery, A. Shihab-Eldin and D. Bodansky. This work was supported in part by a gift to Harvard University from Stone and Webster Engineering Corporation.

References

Arrhenius, S. (1986). *Philos. Mag.* **41**, 237.

Burwell, C.C.; W.D. Devine and D.L. Phung (1983). "Electric Home Heating; it's better than you think," *Energy* **7**, 961.

Fourier, J.B. (1839). *Mem. Acad. Sci. Inst. Fr.* **7**, 569.

Geller, H.S. (1983). *Energy Efficient Appliances*, American Council for an energy-efficient economy and the Energy Conservation Coalition, June.

Goldemberg, J.; et al. (1985). "An end use oriented global energy strategy," *Ann. Rev. of Energy* **10**, 613.

Hogan, W.W. (1979). *Dimensions of Energy Demand*, Discussion paper series E79-02, Energy and Environmental Policy Center, Harvard University.

Hottel, H.C. and J. B. Howard (1972). *New Energy Technologies, Some Facts and Assessments*, MIT Press, Cambridge, MA., 384 pp.

Hyman, M. (1977). "Monitoring solar hot water," *Solar Age*, January and November.

Hyman, M. (1978). "Monitoring Solar Space Heat," *ASHRAE Journal*, March.

IIASA (1981). *Energy in a Finite World—A Global System Analysis*, Ballinger Publishing.

Legasov, V.A. (1986). Presentations at Post-accident review meeting on the Chernobyl nuclear power plant. International Atomic Energy Agency, Vienna, August.

Linden, H.R. (1987). "Electrification of energy supply— an essay on its history, economic impact and likely future," *Energy Sources* **10**, 17.

Marland, (1982). In *CO_2 Review*. W.C. Clark, ed., Institute for Energy Analysis, ORAU, Oxford.

Nordhaus, W.D. (1980). *Thinking about carbon dioxide. Theoretical and empirical aspects of optimum control strategies*, Cowles Foundation paper 565, Yale University.

Nordhaus, W.D. and G.W. Yoke (1982). "Future paths of energy and carbon dioxide emissions" in report of National Academy of Sciences on Carbon Dioxide.

Sakharov, A. (1978). *Bull. of the Atomic Scientist*, **34**, 12 June.

Sakharov, A. (1987). Speech at Forum on a Nuclear Free World, Moscow, Feb. 14-16.

Sakharov, A. (1988). First press conference on first visit to the U.S., Boston, MA, Nov. 6.

Schumacher, E. (1973). *Small is Beautiful*. New York Harper and Row.

Starr, C. (1980). "Energy Futures for the U.S.A." in *Energy for the Year 2000*, R. Wilson (ed.), New York: Plenum.

Stuart, C. (1976). Presentation to U.S. Congress and more recent update using National Petroleum Council data.

Tansil, J. and J.C. Moyers (1974). "Residential demand for electricity" in *Energy Demand: Conservation and Institutional Problems*, S. MacCrakis (ed.), MIT Press.

Uhrig, R.E. (1986). "Toward the next generation of nuclear power plants," *Forum for the Appl. Res. and Public Policy*, Fall, page 20.

U.S. Energy Information Administration (1982).

WEC (1983). *Energy 2000-2020 World Prospects and Regional Stresses*, J.R. Frisch (ed.), London: Graham and Trotman.

GLOBAL CLIMATE CHANGE LINKAGES: STRATEGIES FOR CONTROL OF EMISSIONS

Ian M. Torrens

Electric Power Research Institute
3412 Hillview Avenue
Palo Alto, CA 94303

Abstract

Linkages abound in atmospheric phenomena as well as in methods of dealing with gaseous emissions which lead to these phenomena. This paper explores technologies and methods of reducing emissions of greenhouse and acidic gases, with emphasis on electricity generation. It highlights the need for increasing energy efficiencies and the implications, both within the developed world and for interactions with developing countries, which may be the source of most emissions growth over future decades.

I. The Human Race and Global Climate: Influence Without Control?

The biosphere abounds with linkages. They determine the peskiness of our garden pests, the seriousness of next year's flu epidemic, and whether or not we will have increasingly frequent replays of this summer's Midwest drought in the United States. The earth's partially hospitable land surface has demonstrated a remarkable degree of equilibrium in its interaction with the prevailing climate when measured in centuries or even in a sizeable number of millenia, and when human intervention was a very insignificant perturbation. But past ice ages which bore no relation to human activity remind us that the equilibrium is neither natural nor immutable in the long run, and the fact that we are perturbing the global environment at a rate which is likely to challenge the recent relative climatic stability must give us pause.

The "econosphere," or global economy, is also about linkages. Like the biosphere, it contains a bewildering number of complex feedback mechanisms. Unlike global climate, its natural equilibrium is fragile and short-lived, and its cycles relatively short. But economic cycles notwithstanding, the continued long-term sustainable development of the world economy will be extremely vulnerable to climatic change of a magnitude and rate which impact on the ecosystem's adaptive capabilities.

Copyright 1989 by Elsevier Science Publishing Company, Inc.
Global Climate Change Linkages
James C. White, Editor

Real or potential linkages between economics and climate are discernible not only at the macro level. Their complexity often defies predictive analysis, but not always. Indeed, some such linkages are deceptively simple and revealing. When OPEC countries quadrupled the price of crude oil in the fall of 1973, they set in motion a process which altered drastically the once sacred one-to-one link between economic growth and the growth in total primary energy consumption and, in doing so, sharply reduced the growth rate in CO_2 emissions attributable to fossil fuel combustion in the developed world. How did a short-term oil supply shortage combined within political events in the Mideast result in postponing the year of atmospheric doubling of CO_2 concentrations perhaps by decades? The power of the much maligned economic concept of elasticity of demand was clearly demonstrated, as was the potential for substantial and rapid gains in the efficiency with which energy is converted and used. It is the latter which still today provides us with the best and cheapest way to mitigate global climate change as far as CO_2 is concerned, enabling us to get more gross world product for the same energy input, through more efficient energy conversion and utilization technologies and methods.

It is not the purpose of this paper to dwell on the science of the greenhouse effect, other than to note that, a hot summer notwithstanding, little else has changed over the past year or two concerning the particular phenomenon of climatic warming to justify either the relatively low policy priority given to the issue in previous years or the sudden political consternation which surrounds it today. However, it has joined a number of other global or regional environmental issues which together are raising public awareness of the human race's potential to transform its environment in unintended ways, and are raising the specter of unsustainability of the path we are following. The sardonic headline "Man bites dog" is being replaced by the equally catchy one "Planet bites man," implying that we cannot continue to get away with our historical assumption that the earth's absorptive capacity and ecological resilience are essentially without limit. The human race has perhaps achieved the power to influence its environment on a global scale through its normal activities, on top of the power it already had to do so through nuclear war. It has been successful to date in controlling the latter, but the ability to control the influence of our everyday activities may prove more difficult to achieve.

II. Gaseous Pollutants and Atmospheric Phenomena: Nature of the Linkages

As the complexity of the acid rain/air pollution/climatic change issues continues to grow – with, in particular, different gaseous precursors being implicated in different atmospheric phenomena – there is an increasing tendency for the political debate to draw linkages, however tenuous, among the various global or regional atmospheric phenomena traceable to gaseous emissions in general. Figure 1 tries to bring out some of the relationships in a first-order way. This also attempts to highlight the role electricity generation may play, via gaseous emissions, in one or another atmospheric phenomenon. While this picture may appear complicated, it is undoubtedly a gross over-simplification of the linkages which occur in the atmosphere in reality, and in the interactions between the atmosphere and the rest of the ecosphere. An example of indirect linkages could be increased low-level ozone formation caused by increased ultraviolet solar penetration which in turn is due to stratospheric ozone depletion. Further examples are cited below, in the context of feedback loops.

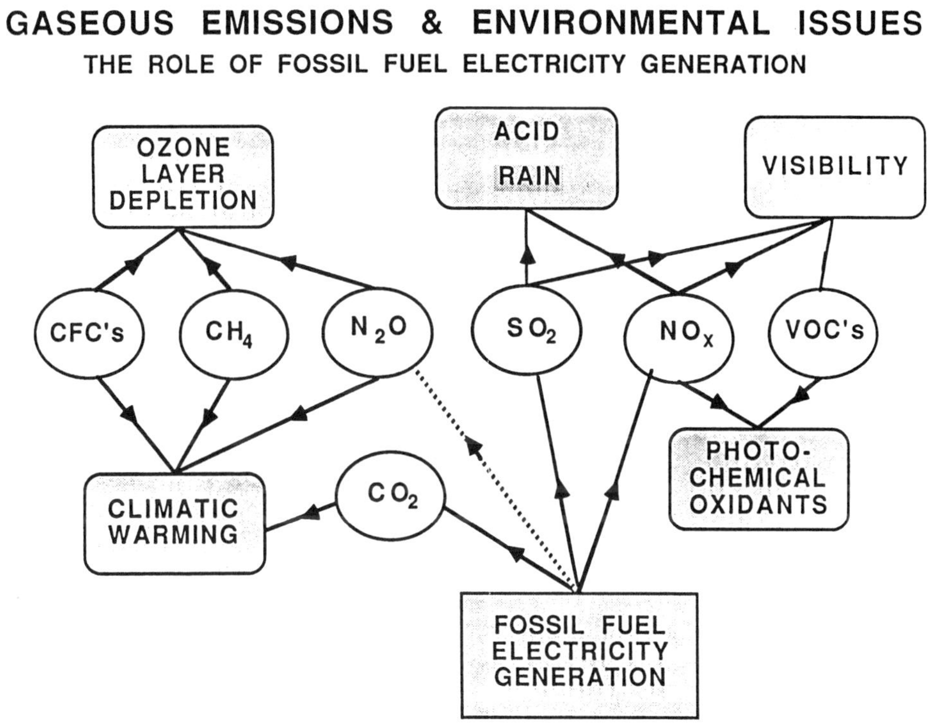

Figure 1

III. Feedback Loops and Climatic Change

The most potentially serious type of linkage is the positive feedback loop. Clear burning of tropical forests adds to atmospheric CO_2 concentrations both by direct combustion and by reducing the earth's CO_2 absorptive capacity. By accelerating greenhouse warming, it may also move climatic zones to higher latitudes faster than temperate forests can keep up, resulting in further deforestation, unintended this time.

Another positive feedback loop, though more speculative, is the loss of reflectivity of the earth's surface due to polar ice melting, leading again to accelerated warming.

A positive feedback loop which directly concerns the U.S. utility industry is the increase in peak cooling demand which would occur if summers become hotter. A corresponding lower heating demand in winter would fall less on electricity than on direct oil or gas consumption, so utilities would face both higher overall demand and sharper peaks in the event of climatic warming. If the response is increased fossil fuel use, the result would be increased CO_2 emissions. EPRI is actively studying the implications of this feedback loop for utility generation planning. (See Section VIII.)

Researchers have also postulated negative feedback loops. Perhaps the most well-known, though still uncertain, is the possibility that a warmer climate will lead to increased cloud cover on the global aggregate, increasing reflectivity and tending to cause some cooling.

It is important to assess these and other possible feedback loops not only as a means of achieving a better understanding of the nature and seriousness of the phenomena, but also in order to develop the most appropriate mitigation strategies.

IV. Origins of Atmospheric Pollutants

Figure 2 shows in schematic form the relative contributions of the different greenhouse gases towards a global climate warming of 1.5°C. It is vital to understand what are

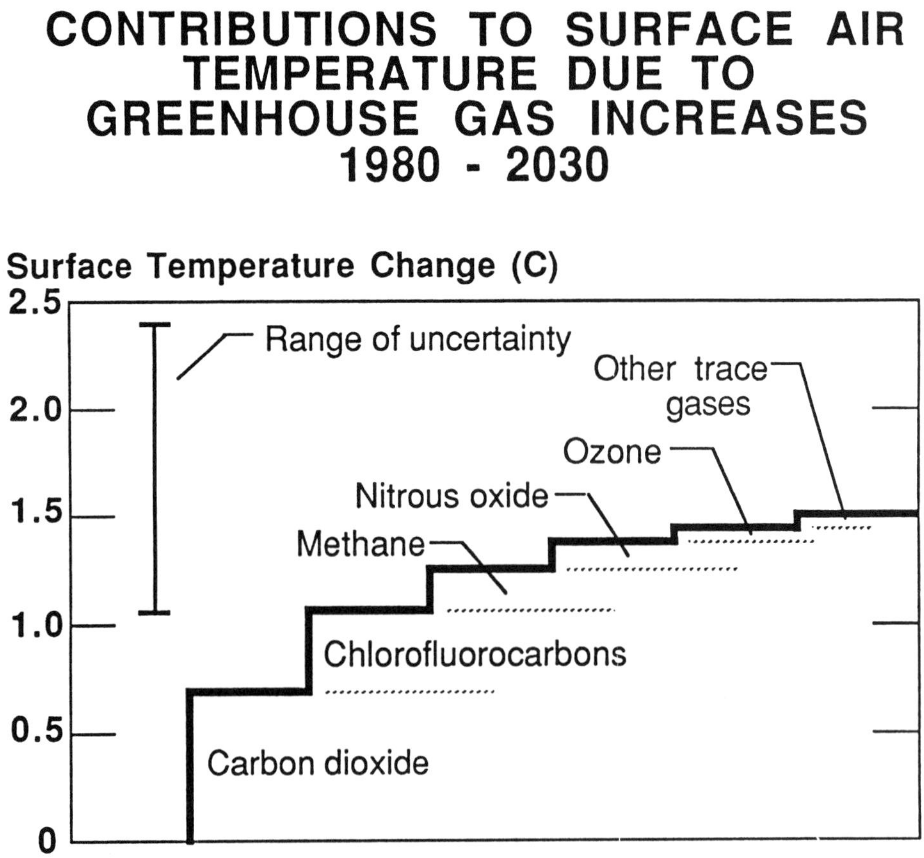

Figure 2

the natural or anthropogenic origins of the gaseous emissions implicated in the various atmospheric phenomena, if we are to develop the most effective strategies to deal with them.

CO_2

Carbon dioxide is thought to be responsible for half of projected global warming. The U.S. produces about 22% of global CO_2 emissions, in approximately equal proportions from industrial processes, electric power production, and transportation, with homes and businesses playing a somewhat smaller role. Electric utilities produce about 35% of U.S. emissions (Figure 3), or less than 8% of the man-made global total and less than that estimated to have been caused last year by burning of tropical forests. The potential for reducing global CO_2 emissions through switching to non-fossil generation in the U.S. utility industry is therefore very limited.

Worldwide man-made emissions of CO_2 more than tripled between 1950 and 1980, and the pattern of global contribution has been changing in parallel with the growth of

**U.S. SOURCES OF CO_2
1987**

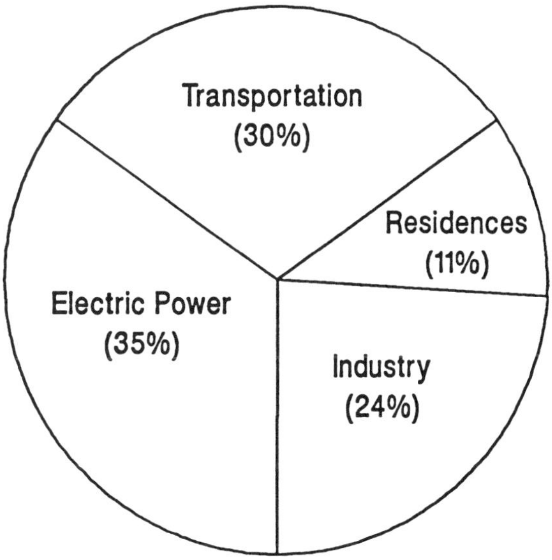

Source: World Resources Institute, 1988

Figure 3

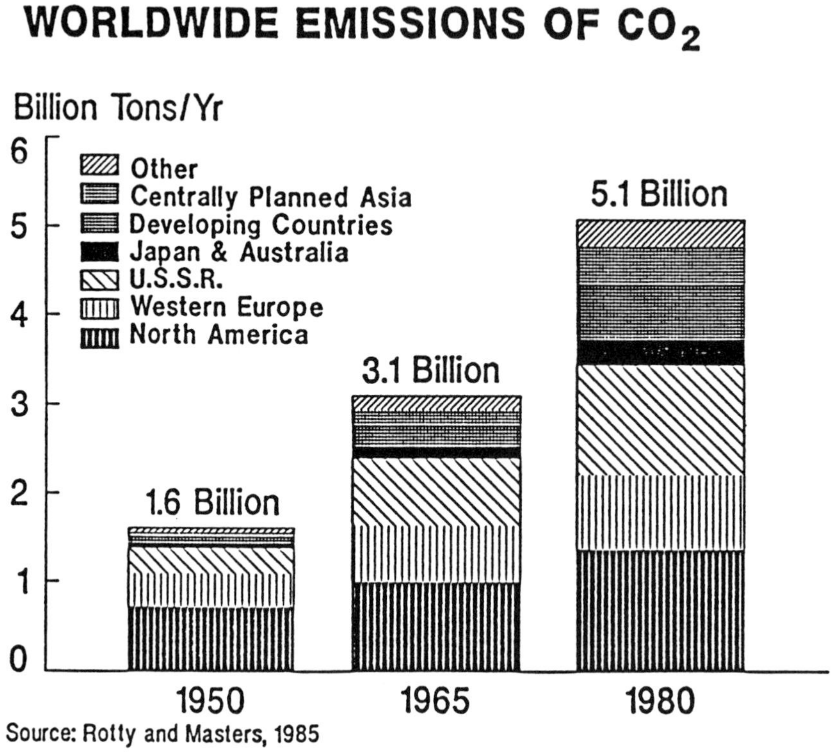

Figure 4

industrialization (Figure 4). North America and western Europe, together accounting for 68% of such CO_2 emissions in 1950, represented only about 43% of the expanded total 30 years later. In contrast, the portion attributable to the People's Republic of China and developing countries in Latin America, Africa, and southern Asia grew from 7% to over 20% of the world total.

Given today's trends, we expect these lesser-developed nations may, according to one projection scenario (Rotty and Masters, 1985), produce two-thirds of the world's man-made CO_2 emissions by the middle of the next century (Figure 5). Given their population growth and their need for industrialization, their desire to control emissions will be very limited without the help of the industrial nations.

With regard to the contribution of electricity generation, if we discount the energy it takes to build an electricity generating plant and manufacture the fuel, and we simply look at emissions in pounds per kWh generated from a variety of fuels, we see a spread from zero for nuclear and renewable sources to about 2 pounds per kWh for a pulverized coal plant; this increases by some 3% when using a flue gas desulfurization system. Advanced technologies – including combined cycle and cogeneration technologies – will

CONTRIBUTION TO GLOBAL MAN-MADE CO_2 EMISSIONS

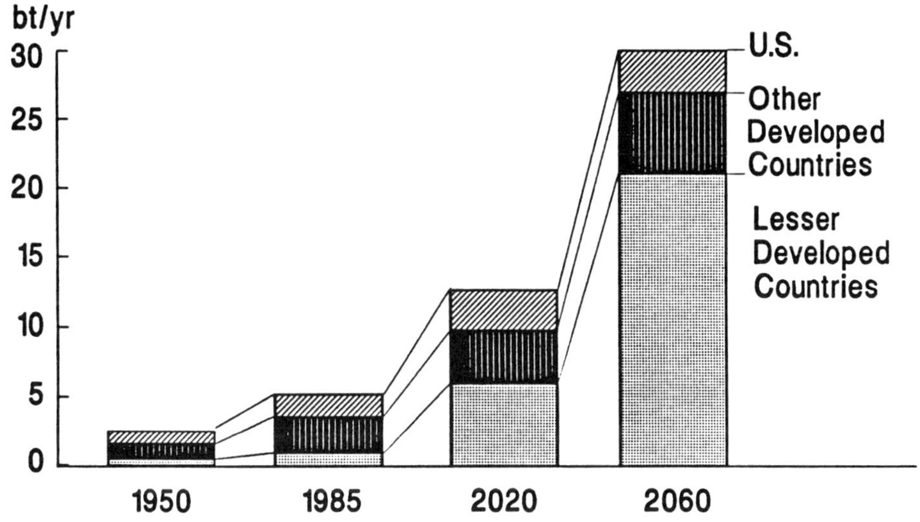

Figure 5

reduce the amount of fossil fuel that is needed to generate a kWh of electricity. Combined cycle technology for example, using natural gas as a fuel, produces electricity with about 45% efficiency as compared to 30-34% for conventional fossil fuel boiler systems, with the added environmental benefits of reduced SO_2 and NO_x. Combined cycle generation with coal gasification (IGCC) is about 38% efficient.

Man-made CO_2, it should be remembered, is only about 5% of the total CO_2 in the atmosphere. Consequently, even though it is increasing, a small change in the natural atmospheric cycle could have a greater impact than CO_2 emissions from human activities. This would in particular affect the conventional climatic model forecasts.

N_2O

This gas is estimated to be responsible for some 8% of the greenhouse effect. Its impact is greater on stratospheric ozone destruction because of its long residence time (150 years): it is estimated to be responsible for 10-20% of ozone depletion. N_2O concentrations in the atmosphere are growing at about 0.2% per year (about half the rate of CO_2). Among natural sources are microbial nitrification, denitrification in soils and ocean surface waters, lightning and natural forest fires. Anthropogenic sources include fertilizer use, biomass burning (e.g., tropical forest clear-burning) and fossil fuel use. Steps taken to reduce deforestation in order to reduce CO_2 growth would also reduce N_2O growth.

Fertilizer use to increase crop yields may be the largest source of atmospheric N_2O growth. This raises a potentially serious linkage problem. While food surpluses persist in the developed countries, reducing excessive fertilizer use would contribute towards mitigating other pollution problems (e.g., nitrates in underground water). However, population growth in the developing countries will inevitably increase pressure to extend fertilizer use in order to increase crop yields to feed the growing population.

Combustion of fossil fuels is also a source of N_2O emissions. Until recently, coal-fired boilers were thought to be the largest anthropogenic source of N_2O, with emissions up to several hundred ppm from each boiler. However, the data supporting this hypothesis were few and inconsistent. To acquire more credible data, EPRI has been developing a continuous N_2O monitor to replace the sampling techniques used hitherto. In doing this, we found an artifact in the sample measurements made at sites burning sulfur bearing fuels. During sample storage, water, NO and SO_2 reacted to form N_2O, producing several hundred ppm N_2O in 2 to 4 hours! The discovery of this artifact also helped explain the lower N_2O emissions measured by several European researchers using other techniques.

Analysis of laboratory and limited field data are now increasingly leading us to believe that N_2O emissions from PC boilers are < 10 ppm. Recent parallel tests by EPA are lending further support to this conclusion. EPRI recently started a field test program sponsored by Southern California Edison to measure emissions from a variety of boilers using our continuous monitor, including oil and gas-fired ones, FBC, combustion turbines, and the cool water coal gasification plant. This series of tests will be completed by the end of 1988, giving us what we hope will be a more definitive picture of this issue. The remaining question – if not combustion, then what is producing the increase in tropospheric N_2O – is being reviewed by EPRI's Environment Division.

CFCs and Methane

CFCs and methane are estimated to be together responsible for about a quarter of potential global climate warming.

With annual growth rate to the present being around 15-20% and their role in stratospheric ozone destruction increasingly confirmed by recent measurements, CFCs were a prime target for control. They are now the subject of an international convention and protocol, and pressure is growing for a still more rapid phaseout of the more harmful varieties. Research and development on alternatives is increasing rapidly, and there are good prospects for substitution in most CFC uses, albeit at a cost.

Methane emissions are rising globally at an estimated 1% annual rate. Sources are estimated to be varied, including rice paddies and swamps, biomass burning (forest fires) and ruminants. With agriculture being the major source, reducing methane emissions runs into the same kind of population pressure issue as N_2O.

Among energy-related sources of methane are the natural gas fuel cycle, coal mining and oil production. Increased gas prices could in the future provide an economic incentive to reduce these methane emissions. Also, underground coal gasification is a technology which could, if its economics improve, contribute to reducing methane emissions associated with coal mining.

Methane is also produced by anaerobic fermentation from landfills of urban waste, and is in some places captured and piped into gas pipelines. Again, the economics are quite marginal at present gas prices.

Methane turns out to be linked to another energy related gas, carbon monoxide, which plays a role in the overall chemistry of atmospheric greenhouse gases. Its increase over the industrial era has acted to maintain atmospheric methane concentrations higher than if it were absent. This is yet another synergism between global and more local atmospheric issues.

SO_2

Power plants are clearly the major sources of SO_2 emissions in most industrialized countries (Figure 6). Figure 7 shows that in the United States, while SO_2 emissions have been decreasing, utility emissions are still responsible for some two-thirds of the total emissions of this gas. Power plants are consequently prime targets for SO_2 emission reductions.

NO_x

The main source of NO_x is the combustion process. For this gas, implicated in both acid rain and photochemical oxidant formation, emissions have been on a growth trend in the United States and most other industrialized countries. Mobile sources constitute the main source in the United States and most countries; the utility industry is responsible for approximately one-third of total U.S. emissions (Figure 6). This fact does not, however, remove the power industry from the front line as far as NO_x control is concerned, since it is easier to target large single emitting sources than a multitude of underperforming mobile sources.

The United States has joined a new international protocol under the UN Economic Commission for Europe's Convention on Long Range Transboundary Air Pollution, which provides for a standstill in NO_x emissions.

V. Controlling SO_2 and NO_x Emissions

A consequence of high costs for environmental control in new power plants, coupled with the continuing possibility of new emission reduction regulations affecting existing plants, has been the development over the middle years of this decade of the concept of *clean coal technologies* (CCT). By improving the effectiveness of control technologies and reducing their costs, this concept seeks to reduce the policy conflict between coal use and the environment. Clean coal technologies have evolved into a family of precombustion, combustion and postcombustion technologies designed to provide the coal user

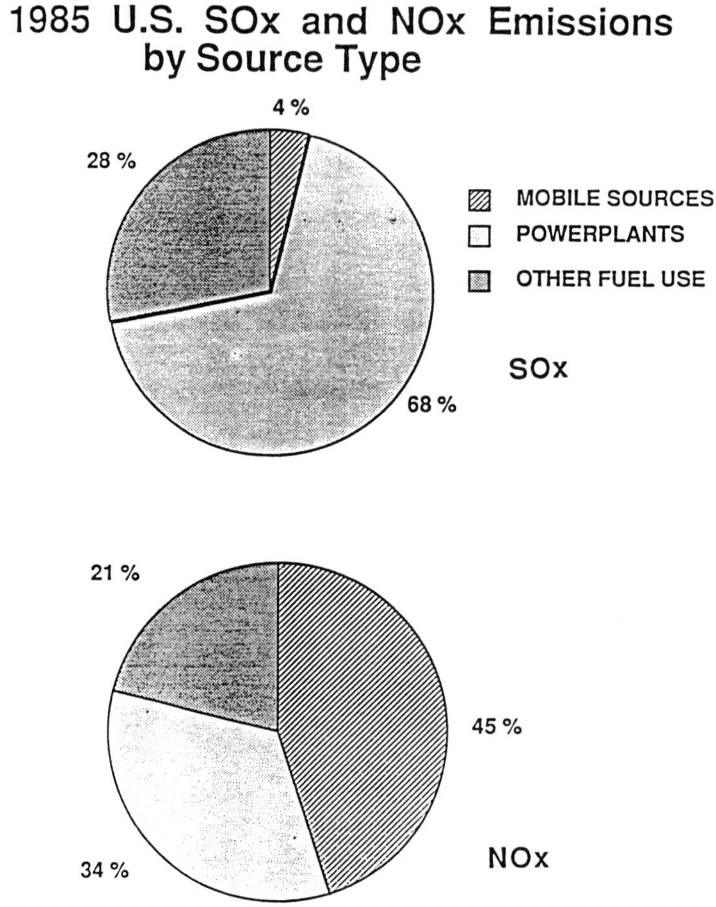

Figure 6

with added technical capabilities and flexibility at lower cost in terms of environmental control.

The Department of Energy in February, 1988, issued its second Clean Coal Technology solicitation. This new program is offering funds totaling $537 million for innovative CCT projects where industry will put up at least matching funds. A total of 16 projects were selected by DOE, with the accent heavily on innovative retrofit SO_2 and NO_x control technologies.

The DOE selection clearly recognized the fact that clean coal technologies include postcombustion retrofit technologies like FGD. There is no need – in fact, it is inappropriate – to exclude *improved* commercial retrofit control technologies from the CCT menu. When it comes down to the investment decision point, these are the technologies against which the performance and economics of developing and innovative CCT's like AFBC, CGCC and other NO_x or SO_2 reduction techniques, now emerging through pilot

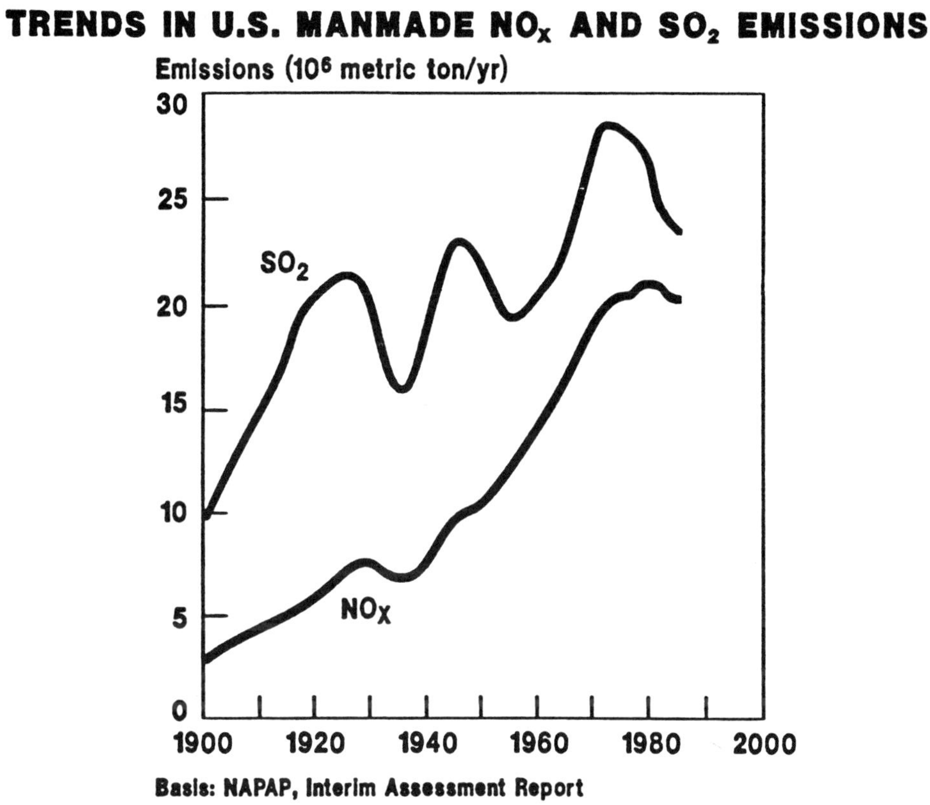

Figure 7

and demonstration stages towards deployment, will be judged. This benchmark is a constantly moving target, since technological improvements are not restricted to specific types of technology. Today's postcombustion control technologies *are* superior to those of the 1970's in reliability, cost, and environmental performance, as recent experience, especially in Europe, is showing.

At the same time, a "level playing field" for CCT means ensuring that *tomorrow's* technologies are given a chance to show their mettle and compete against today's. Some legislative proposals before Congress aimed at acid rain control, involving large power plant retrofit programs, would definitely tilt the playing field away from innovative clean coal technologies, probably achieving short-term air pollutant emission reductions at the expense of lengthy delays in developing the fully integrated, environmentally benign coal processing technology of the 21st century.

The net cumulative effect of these delays would probably be negative on our environment, because *advanced coal processing* offers important improvements in efficiency of energy use, in resource recovery with waste minimization, and even in potential reduction of CO_2 emissions, in addition to virtual elimination of SO_2 and NO_x emissions. The "*coal refinery*" concept integrates coal conversion (gasification or liquefaction) with cogeneration to produce process steam and electricity, as well as feedstocks for chemical processing, and liquid or gaseous fuels. It is even conceivable that coal refining could be further integrated with the high temperature nuclear reactor which could supply 1,000°C process heat for coal processing and gaseous fuel production (including hydrogen). Regulatory initiatives should preferably encourage, but at least not hinder, progress toward such longer-term visions of clean coal use.

The development of clean coal technologies is proceeding apace, and will likely accelerate as a result of the cooperative government-industry demonstration program being carried out through DOE. Given the needed regulatory flexibility in any future emission legislation, there are good prospects for bringing a broad array of clean coal technologies to commercialization over the next decade, as well as positioning the United States as a technological market leader in clean coal use. Figure 8 gives an approximate timetable for developing some principal clean coal technologies, as far as can be seen at present.

ESTIMATED COMMERCIAL AVAILABILITY DATES OF ICCTs

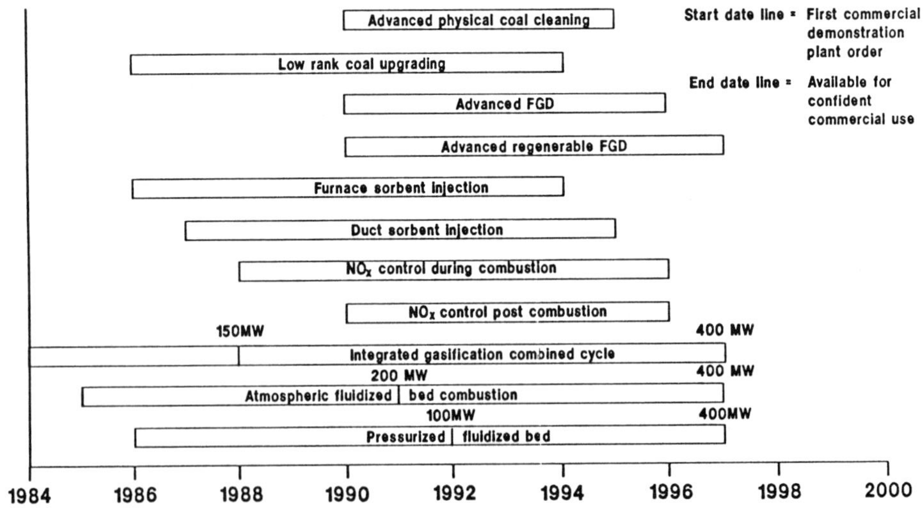

Figure 8

VI. Options to Reduce CO_2 Emissions from Power Generation

Table 1 and Figure 9 compare the CO_2 emissions per kilowatt-hour of electricity generated for different types of power generation, including both conventional and advanced (not yet commercial) generation cycles. Coal is predominant in the U.S. electric power sector accounting for more than half of electricity generated (Figure 10). The hydrogen-to-carbon ratio in coal puts it at a disadvantage compared to oil and natural gas. The differences in the coal generating systems depicted here are due to differences in generation efficiency and to the use of alkaline sorbent for sulfur removal (FGD and AFBC).

To reduce CO_2 emissions from power plants, five broad options have been suggested. These are:

- Increase efficiency of primary generation using improved techniques in electronic controls, heat recovery, and environmental control systems, as well as advanced conversion systems such as combined cycles and fuel cells;
- Implement energy conservation schemes and improve end use energy efficiencies;
- Shift to high hydrogen content (per carbon mass) fuels (natural gas preferred over coal); use hydrogen as a primary fuel for power generation and transportation (the hydrogen being generated by non-fossil means – see below);
- Remove CO_2 from effluents by end-of-stack acid gas removal techniques;
- Shift to non-fossil fuel electricity production, through renewable and nuclear alternatives, including eventually fusion;
- Another option which does not reduce power plant emissions, but contributes toward overall CO_2 emission reductions, is cogeneration with waste heat utilization.

Increased Generating Efficiency

Current methods to convert thermal to electrical energy have an efficiency of about 35%. This can be increased with new technology to approach 50% efficiency; with combined cycle methods and fuel cells, efficiencies of 60% are believed to be achievable. Thus, the same amount of electricity could be produced with almost half as much fossil fuel consumption. The cost of electricity would rise somewhat if the technology were adopted, primarily from capital investment. However, the increase would be substantially less than that which would take place if technological options for CO_2 removal were required. Also, more modest incremental increases in generation efficiency have proportionately larger impact on CO_2 reduction, because of the multiplier effect (e.g., 5 percentage point increase in efficiency reduces CO_2 emissions by about 15%).

TABLE 1

CO_2 EMISSIONS IN LBS/KWH FOR DIFFERENT FUELS AND TECHNOLOGIES

Technology	Wt. Fraction Carbon in Fuel	Fuel BTU/LB	Overall System Heat Rate BTU/KWH	Stoich Ratio CaCO3	LBS/CO2 Per KWH
PC with FGD	0.575	10,100	9,850	0.9	2.10
AFBC (Circulating)	0.575	10,100	9,750	2.0	2.14
PFBC	0.575	10,100	8,710	2.0	1.91
Oil-Steam #2	0.874	19,400	9,400	0.0	1.55
IGCC	0.575	10,100	9,010	0.0	1.88
Coal Gasif Molten Carbonate Fuel Cell	0.575	10,100	7,130	0.0	1.49
Nat. Gas Comb. Turbine	0.750	23,500	11,000	0.0	1.29
Nat. Gas Comb Cycle	0.750	23,500	7,900	0.0	0.92

Notes:
1. Performance and fuel data from EPRI TAG (1986).
2. For the PC plant with scrubber, stoichiometric patio assumes 90% conversion (same as sulfur removal percentage)
3. For both fluidized beds. The desulfurization is conducted with limestone.

SOURCE: M. Gluckman (EPRI): Private communication.

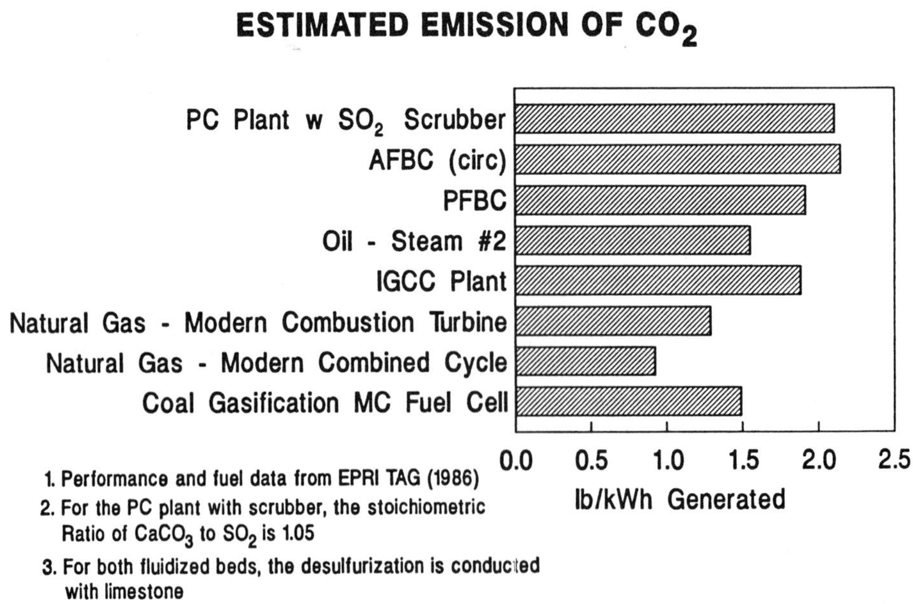

Figure 9

ELECTRICITY GENERATION BY FUEL TYPE (1987)

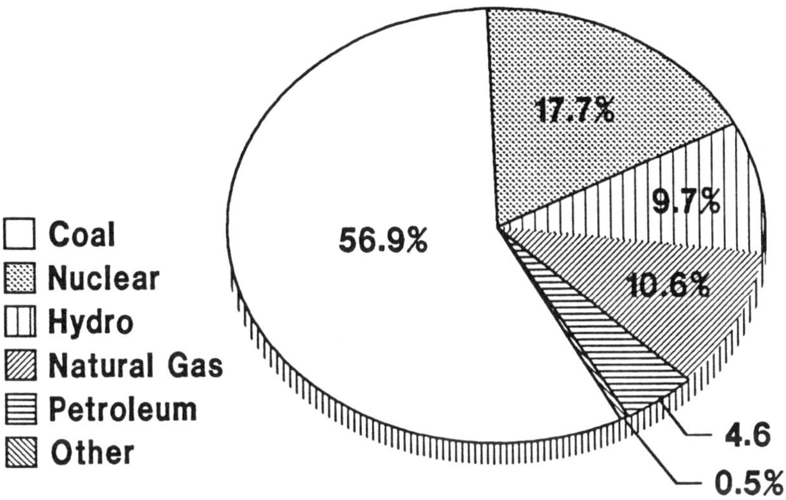

Source: Edison Electric Institute

Figure 10

End-Use Efficiency

Energy end-use efficiency gains are particularly attractive in developed countries where electrical user efficiency improvements can be applied relatively easily. Significant opportunities in the electricity sector include the development and application of energy-efficient electrical machinery in industry, improved electronic controls in heating and air-conditioning use, and improved lighting efficiency.

Although historically the health of the electricity industry is based in part on load growth, the industry has taken major steps to shift the emphasis. Demand side planning, conservation and increased efficiency in electrical use are areas for major utility investment in the future.

Shifting Fuels

A shift to more hydrogen-rich fuels would involve much greater use of natural gas than at present. The use of natural gas would tend to reduce CO_2 emissions per unit of energy produced relative to coal or petroleum.

The main barrier to increased natural gas use, especially for electricity generation, is its estimated reserve base, only a few decades compared to centuries of reserves for coal. Natural gas prices are low now, but are expected to increase in time with rising demand.

Biomass consumption (burning) is sometimes cited as a means of reduction in CO_2 effluent growth. In this scenario, there is no net increase in CO_2 levels over long periods of time. Unfortunately, this proposal does not account for major short-term time differences in biospheric assimilation of CO_2 by photosynthesis and the combustion process. These differences may or may not be critical to atmospheric CO_2 changes. Land use issues are also critical in implementing a biomass based fuel program. For example, the Four Corners power plants in Arizona and New Mexico (2000 MW) have been estimated to require over 3000 mi^2 of forest—an area about one-third the size of Massachusetts if they were replaced by biomass (Oman, 1988).

CO_2 Scrubbing

The recovery of CO_2 from power plant effluents has been suggested as a fourth alternative. Flue gas usually contains 12-15% carbon dioxide and is at atmospheric pressure. It can be recovered with existing scrubbing technology, for example, using chemically alkaline absorbent systems, followed by liquefaction and transmission to eventual disposal or reuse.

Several studies have evaluated potential markets and uses for recovered carbon dioxide. In Germany, for example, at a coal gasification plant that produces alcohol, carbon dioxide is recovered and used for food processing.

Under certain conditions, the recovery of oil can be increased by carbon dioxide injection. During the 1970s, a power plant in the southwestern U.S. was used as a source to supply carbon dioxide for an oil field. Sulfur oxides were not scrubbed, and it was concluded that the use of impure CO_2 had no effect on the efficiency of enhanced oil recovery. However, it is unlikely that carbon dioxide containing sulfur dioxide could be widely used for enhanced oil recovery since only about 50% stays "down-well."

Steinberg and Cheng (1988) estimated that recovering carbon dioxide and its transportation would add 35% to 80% to the electricity costs, depending on the type of power plant and other factors. Figure 11 extrapolates Steinberg and Cheng's data to estimate the cost in 1987 dollars of scrubbing CO_2 from all United States fossil fuel power plants liquefying it, piping it to and sequestering it in the deep ocean, and replacing with nuclear plants the 14% power lost due to the energy demand of this process. However, since these estimates concern largely untried technologies, the costs could be substantially higher. Whatever the real cost, CO_2 recovery for power plants seems very expensive considering the total reduction in CO_2 to be achieved based on the percentage contribution of electricity generation to total emissions of CO_2. This is especially relevant considering that the final fate of the CO_2 recovered is uncertain, and there is no guarantee that it will not eventually find its way into the atmosphere.

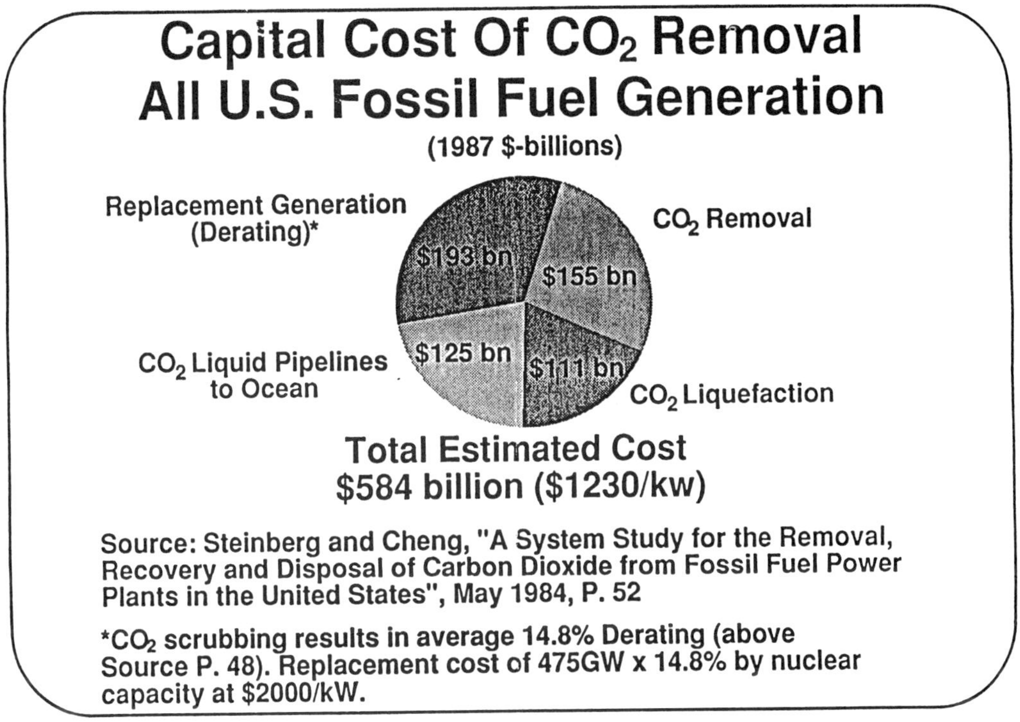

Figure 11

Switching to Non-Fossil Energy Systems

The last option of non-fossil fuel energy focuses on hydroelectric, nuclear generation, and solar. If the current public opposition to nuclear power could be changed and new nuclear plants could be built, the current projected cost of nuclear electricity could be reduced to a level competitive with coal-fired generation. This would be less than a fossil fuel system with CO_2 recovery.

The opportunities for solar energy production are improving with advances in photovoltaic devices. However, this process remains more costly than conventional electricity by a wide margin (e.g., estimates exceed 25¢/kWh). Projected technological improvements for the mid-1990's in photovoltaic conversion efficiency could bring the cost of solar electricity below 10¢/kWh. Solar energy use requires storage capability, and will vary geographically in efficiency. At best, this choice will be a necessary complement to other methods, but is unlikely to be a dominant energy resource.

Simple and direct linkages exist among strategies. Most of the above CO_2 mitigation strategies would provide the ancillary benefit of reducing emissions of other gaseous pollutants, principally SO_2 and NO_x. Efficiency improvements would moreover pay back investments in reduced cost of energy services. Indeed, all of these consequences, independently or collectively, are sound reasons to push ahead with the efficiency strategy, both in the industrialized countries and in those who need to allocate scarce capital resources to their future development in the most efficient and cost-effective way possible. (See Section X.)

Less simple and less direct linkages among strategies also exist, probably in abundance. We are only beginning to explore them. The next section considers the role that the electric vehicle might play. But it is clear that the linkages extend far beyond the electricity sector, indeed, far beyond the energy sector. For example, a policy to move towards a society which creates less solid and liquid wastes would in all likelihood produce benefits in terms of reduced atmospheric pollutant emissions. A major assessment effort is needed to identify, quantify and draw conclusions from these linkages.

VII. Electric Vehicle as an Emissions Reduction Strategy

The internal combustion engine, besides being an ubiquitous emitter of several gaseous pollutants, is less than 20% efficient in conversion of primary energy into useful energy service. Some actual gains in overall efficiency may therefore be realizable by replacing internal combustion vehicles by electric vehicles, with electricity provided by an efficient power plant. If the power were non-fossil generated, the reduction in vehicular gaseous emissions would be total. If fossil-generated, this would depend on the fuel used and (for emission other than CO_2), the pollution controls on the power plant. An ancillary benefit would be the ability to use off-peak electricity to charge the electric vehicle's batteries. In areas where there is nuclear generation, most off-peak electricity would be baseload, and therefore would tend to be nuclear.

Advances in battery technology are improving the economics of electric vehicles. Urban pollution problems in some areas already provide an argument for developing the electric vehicle option through an incentive approach.

The question has been raised whether the argument for electric vehicles could be reinforced by the CO_2 emissions issue depending on conditions of generation of electricity. Table 2 compares estimates of CO_2 emissions per mile of an electric versus a gasoline-driven vehicle, assuming the electricity is generated by a coal-fired power plant. The results indicate that CO_2 emissions reduction would not be an argument for electric vehicles if the electricity is generated from coal.

The electric vehicle is one strategy where it would be very worthwhile to assess more fully the linkages which exist in both the energy and gaseous emissions fields.

VIII. Adapting to Warming: The Utility Case

If the earth's climate does change markedly in the next century, societies and institutions will have to adapt. Foresight and planning can help make this adaptation easier. Section III discussed linkages which take the form of feedback loops, including one which could directly affect electric utilities. EPRI took a first step in this direction recently by

TABLE 2

ELECTRIC VEHICLES AND CO_2 EMISSIONS

1.	ELECTRIC VEHICLE	CO2 Emissions (Lbs/mile)
	Present Consumption (1 mile/kWh)	2.31
	Projected 1995 Consumption (2 miles/kWh)	1.16

Assumptions

Pulverized Coal Power Plant Emissions:	2.1 lbs/kWh CO_2
Transmission and Distribution Loss:	0.21 lbs/kWh CO_2 (10%)
Total Emissions:	2.31 lbs/kWh CO_2

2.	CONVENTION GASOLINE DRIVEN VEHICLE	CO2 Emissions (Lbs/mile)
	20 mpg	1.07
	25 mpg	0.86
	30 mph	0.72

Assumptions

Density 6.5 lbs/gallon, 90% carbon

CO_2 Emissions $= 6.5 \times 0.9 \times \frac{44}{12} = 21.45$ lbs/gal

conducting a preliminary "what if" study to explore how utility operations might be affected if greenhouse-induced climate changes do occur (Hidy, 1988).

The study was conducted by ICF (1987), and cosponsored by EPRI, the Environmental Protection Agency, Edison Electric Institute, and the New York State Energy Research and Development Administration. It concluded that climate change may entail significant capacity and cost implications for utilities. Only two of the possible ways that utilities might be impacted were examined: the effects of projected temperature rises on heating and air-conditioning loads, and the implications of reduced stream flow on hydroelectric resources. There were two case studies: a southeastern utility and the member companies of the New York Power Pool.

Focusing on the year 2015, a point well within utility planning horizons, the analysis suggests that temperature increases of slightly under 1°C would increase the need for new generating capacity in both service territories by 10 to 20%, leading to hundreds of millions of dollars in increased electricity production costs.

The study also looked at the implications of alternative planning assumptions made by the utilities with regard to climate change. If the southeastern utility planned for and phased in new combined-cycle generating capacity in anticipation of a warming that never occurred, it would find itself with excess capacity, costing an estimated $11 million in the study year of 2015. If, however, the utility did not plan for the warming and it did occur, the costs of hurriedly adding relatively expensive short-term supplies like oil-

and gas-fired turbines and purchased power were shown in this analysis to be even more expensive—about $55 million for the year 2015. In this scenario then, the risk of not anticipating the warming is five times as great as the risk of planning for it and having it not occur.

Whether or not the numbers in this scenario are accurate, it is a useful exercise to evaluate in even a qualitative way what might be at stake for the industry. It looked at only two utilities in a single year and considered only two of the ways that climate change could affect utility operations. Integrated across many years, many impacts and many utilities, the cost to the industry and its customers of guessing wrong about climate change could be very high.

Future impact assessments by EPRI will consider more of the pathways by which climate change can affect utilities, such as demographic shifts, changes in pumped irrigation loads, changing storm patterns, and the implications of a rising sea level for coastal plants. The methodology can be improved and the impacts need to be assessed regionally across the country to give a more balanced and comprehensive view.

IX. Reforestation as a Mitigation Strategy

Reforestation has been widely suggested as a mitigation strategy. The fact that new growth is most efficient at absorbing CO_2 suggests that tree farming should be supported, with mature trees used for lumber to store the carbon taken up from the atmosphere.

Marland (1988) estimated that to absorb the 5 billion tons of carbon emitted as CO_2 every year by fossil fuel combustion would require planting 7 million square kilometers of trees—an area the size of Australia, or about as much as all of the tropical forest cleared by human activity up to the present. Reforestation has also been suggested in the U.S. context, in a proposal by the American Forestry Association, to plant 100 million new trees over the next 4 years, thereby absorbing about 18 million tons of carbon from the atmosphere each year.

One U.S. company active in the independent power production field, Applied Energy Systems, has contracted with the World Resources Institute to provide funds assisting in a forestry project in Guatemala as an offset for the CO_2 emitted from a power plant it owns in Connecticut.

The use of wood as a fuel for power generation is another possible way to mitigate CO_2 emissions by substituting it for fossil fuel, though it has limitations (see Section VI).

X.Interdependence and Global Climate: The Technology Transfer Challenge

In an increasingly interdependent global economy, mitigating CO_2 growth also involves linkages which become more evident should climatic warming be confirmed. While the industrialized world can possibly level off or even reduce its CO_2 production due to fossil fuel combustion through efficiency increases and fuel switching, the likelihood is that the developing countries will continue to increase their use of fossil fuels as they move up the growth ladder.

There is neither political feasibility nor probably economic common sense in any attempt to curtail developing countries' growth in fossil fuel use. However, this is an area where developed and developing countries have a clearly discernible common interest in a world subject to global warming, which would not be so obvious absent the concern over climatic change. The developing countries by and large are prevented by economic reasons from having access to state-of-the-art technologies for such vital symbols of development as electric power or industrial plants. For countries where capital is scarce, first cost is always much more important than lifetime cost. They generally have to import technologies of lower energy conversion efficiency, sometimes outmoded technologies when measured by the industrialized world's standards, in order to obtain the lowest initial capital cost available.

In the past, this has been an issue where there was no clear convergence of interests between the developed and developing world. In the future, if climatic warming is confirmed as a global issue, it will be clearly in the industrialized countries' interests to provide developing countries with the ability to minimize the growth of their contribution to atmospheric CO_2. Indeed, there will likely come a time when incremental energy efficiency increases will be much cheaper to obtain in the developing countries.

This suggests that an important part of any strategy to mitigate global climatic warming will be to facilitate transfer of technologies to the developing world – not necessarily the Rolls Royce or Ferrari of technologies, but equally not the 1960's Chevrolet or Volkswagen! One need only consider that if fossil fuel power generation efficiency could be increased in the developing world from the high 20's in percent to the high 30's (still below state-of-the-art capabilities), the same amount of electricity could be produced with three power plants that today is produced with four. CO_2 emissions would fall by a quarter in terms of useful energy produced. Other pollutant emissions would also fall proportionately, yielding ancillary environmental benefits of a more local or regional nature.

Technology transfer begins with information transfer, which could benefit significantly from government-industry cooperation, including the channels of the multilateral development banks and international or national development assistance organizations.

An R&D organization like EPRI possesses an extraordinary resource of technological information which, while most of it is publicly available, does not get widely disseminated beyond its member utilities. While EPRI's primary mission is clearly to its members, their own interests could suggest a more pro-active stance toward cooperation in technology transfer for development in the event that CO_2 mitigation strategies seem to be desirable in the global context. The same clearly applies to other research organizations and to major branches of industry.

XI. Conclusion

Whether or not the global climate is already changing in response to greenhouse gases, whether or not we are already committed to losing a good part of the protective stratospheric ozone layer as a result of human activities, whether or not the effects on the ecosystem of acid rain and photochemical oxidants are such as to require large and urgent acid gas emission reductions of the type undertaken in some countries, the fact is that public consciousness of these issues and through them of the potential vulnerability of "spaceship earth" has grown rapidly over the past 12 months. While the public cannot always understand the nature of global atmospheric linkages—even the experts have much to learn about these—the broad public perception is that these linkages do exist and that the major atmospheric issues should not each be addressed in isolation.

This environmental consciousness is leading to calls on policymakers to react in some way, and the pressure is on to do more than simply further research. The challenge will be to achieve the correct balance between two broad policy thrusts: on the one side R&D which is clearly needed on both the complexities of the phenomena and the mitigation techniques, and on the other side an appropriate mix of the less costly and more effective actions to improve energy efficiency and reduce emissions of the series of greenhouse and acid gases, which are not being taken at present for economic or institutional reasons. The answer is certainly not one *or* the other, and finding out where to strike the appropriate balance will not be easy, in view of the scientific uncertainties.

Furthermore, given the global nature of the problem, the balance will probably not be achievable unless it is sought through a process of international consensus-building. Here it is clear that the western industrialized countries, because they are the greatest energy users and possessors of the technologies (and the capabilities to develop the technologies) to mitigate greenhouse warming and other atmospheric pollution phenomena, will have to take the leadership role, but in doing so will have to be sensitive to the needs and aspirations of the world's large developing populations.

The air we breathe is the ultimate linkage affecting human existence across the globe. Perhaps the potential threat of climatic change induced by greenhouse warming will bring some degrees of warming to international environmental relations, though it is difficult to envisage this so long as the threat is still only perceived as potential and long-term. In any event, it is clear that the linkages involved in global climate change are not only scientific and economic, but political as well. Will this be an issue where the perceived self-interest of countries will converge to a greater extent than it has in the past, where environmental issues with international implications were concerned? Or will we have some cold winters and cooler summers, continued lower oil prices and business-as-usual? Conscious of the limitations of his crystal ball, this author is not going to make any predictions!

References

General

"The Politics of Climate," *EPRI Journal*, June 1988.

Chauncey Starr, "Global Climate Change and the Electric Power Industry," Presentation to the National Climate Program Office Strategic Planning Seminar, Washington, DC (January, 1988).

I. Mintzner, *A Matter of Degrees: The Potential for Controlling the Greenhouse Effect*, World Resources Institute Research Report No. 5, Washington, DC (April 1987).

J. Goldenberg, T. Johansson, A. Neddy and R. William, *Energy for a Sustainable World*, World Resources Institute, Washington, DC (April, 1987).

Jessica Mathews, "Global Climate Change: Toward a Greenhouse Policy," *Issues in Science and Technology*, Vol. 3, No. 3, p. 57 (Spring 1987).

Specifically mentioned in text

ICF, Inc., *Potential Impacts of Climate Change on Electric Utilities*, for EEI, EPRI, NYSERDA and USEB (1987).

Oman, H., "Controlling CO_2 Buildup in the Atmosphere," *Chem. Tech.*, Feb. 1988, p. 116.

Laurmann, J., "Strategic Issues and the CO_2 Environmental Problems" in *Carbon Dioxide: Current Views and Developments in Energy Climate Research*, p. 415, Eds. W. Bach et al., (C. Reidel, NY, 1983).

G. M. Hidy, "Jeckyll vs. Hyde, the Electric Utility Industry and Climate Change," Proceedings of Symposium: *The Projected Global Warming: Policy Issues and Options*, University of Washington (April 1988).

G. Marland, *The Prospect of Solving the CO_2 Problem Through Global Reforestation*, U.S. DOE Report M. DOE/NBB-0082 (February 1988).

R. M. Rotty and C. D. Masters, "Carbon Dioxide from Fossil Fuel Combustion: Trends, Resources, and Technological Implications," in *Atmospheric Carbon Dioxide and the Global Carbon Cycle*, U.S. DOE Report DOE/ER-0239 (December 1985).

M. Steinberg & H.C. Cheng, "Advanced Technologies for Reduced CO_2 Emissions," Proceedings of 1988 APCA Annual Meeting, Dallas, TX (June 1988).

Climate, Air Quality, Energy Alternatives and Efficiency

William R. Moomaw

World Resources Institute
1709 New York Avenue, NW
Washington, DC

There is a sense that we are being overwhelmed by an overabundance of local, regional and global problems. The traditional policy response of waiting for each to become sufficiently serious before it is addressed on an individual basis is no longer a viable approach. Because the atmospheric problems of global climate change induced by an accelerating greenhouse effect, urban and regional air pollution, acid deposition and stratospheric ozone depletion are interlinked both scientifically and economically, it is necessary that proposed solutions recognize this fact and be designed appropriately. Unless this is done, we are not only likely to exacerbate one problem as we attempt to solve another, we may also miss cost-effective options which only become apparent when an integrated policy analysis is undertaken.

It is possible to begin with any one of the atmospheric issues as a starting point and examine its relationship to each of the others. Because it is the most all encompassing, I have chosen in this paper to utilize global climate change as the central issue in this analysis and to relate the other problems to it.

Looked at from the perspective of the infrared heat-trapping gases involved, about 50 percent of the increase in global warming arises from carbon dioxide, approximately 18 percent each from methane and chlorofluorocarbons (CFCs), perhaps 8 percent from tropospheric ozone, 6 percent from nitrous oxide and a small contribution from other sources.[1] Policy is not, however, based upon chemistry, but rather on economic sectors. When seen in this light, three-quarters of the carbon dioxide comes from the energy sector (the remainder is principally from deforestation and agricultural practices), but additional energy sector activities contribute methane, ozone and nitrous oxide, making the energy sector by far the largest single contributor to climate change. CFCs are also heavily used by the energy sector both as refrigerants and as insulators, and play the principal role in stratospheric ozone depletion in addition to being the most rapidly growing greenhouse gases. Some estimates of the contribution to atmospheric warming of each gas by economic sector are summarized in Table 1.

In addition to its role in global warming, the combustion of fossil fuels is also primarily responsible for acid deposition and the urban and regional smog that adversely affects

Copyright 1989 by Elsevier Science Publishing Company, Inc.
Global Climate Change Linkages
James C. White, Editor

Table 1

Science to Policy Translation Matrix for Global Warming from Greenhouse Gases

PERCENT OF GLOBAL WARMING

Gas / Sector	Carbon Dioxide	Methane	Ozone	Nitrous Oxide	Chlorofluorocarbons (and others)	Percent Warming by Sector
Energy Direct	35	*3*		*4*	---	49
Indirect		*1*	*6*			
Deforestation	*10*	*4*	---	---	---	14
Agriculture	*3*	*8*	---	*2*	---	13
Industry	2	---	*2*	---	20*	24*
Natural Feedbacks	†	†	†	†	---	?
Percent Warming by Gas	50	16	8	6	20*	100

Values are estimates of the percent of global warming between 1980 and 2030 arising from each gas and by policy sector, assuming current trends. Numbers italicized and underlined are very uncertain.

*** This assumes current trends for CFCs. If the Montreal Protocol is fully implemented, CFC contributions to warming would drop.**

human health and damages crops and forests. What is not so immediately obvious are some of the strategies that might simultaneously address these problems. Tropospheric ozone arises from atmospheric photochemistry. The ingredients are volatile organic compounds (VOCs) and oxides of nitrogen (NO_X) that arise during combustion in vehicles, electric generating stations and, to a lesser extent, from industrial and domestic heat production [2]. NO_X is also a major precursor of acid deposition [3]. From a global warming perspective, ozone may account for only 8 percent of the problem, but there are several considerations that suggest it may be more important than that.

First, there exist feedback mechanisms that link increased global warming to higher ozone production. This ozone increase is expected to occur partly because of increased electricity production to meet greater demand for air conditioning which can be expected to generate more NO_X and other air pollutants. More important is the strong correlation between ozone production and higher temperatures. It is expected that the number of heat waves is likely to increase sharply in many urban areas [4] even for a relatively small rise in global average temperatures creating ideal conditions for greater smog and ozone formation [5].

Second, unlike other contributors to global warming, once formed, ozone has an atmospheric lifetime of only hours to days, rather than decades to a century [6]. Therefore, eliminating the precursors of ozone brings about the simultaneous and immediate reduction in a greenhouse gas and a source of air pollution and acid rain. In other words, a conventional clean air strategy can have benefits for climate change as well. Not all clean air strategies will, however, achieve these simultaneous benefits. For example, the proposal to replace gasoline with methanol to reduce conventional air pollutants will produce large quantities of extra carbon dioxide if that methanol is produced from coal.

A similar argument can be made regarding the importance of considering methane controls in multiple contexts. The natural atmospheric sink for methane is its reaction with hydroxyl radical [7]. Because large quantities of a non-greenhouse gas, carbon monoxide, that come largely from incomplete combustion in automobiles, compete for the highly reactive hydroxyl radicals, concentrations of methane are higher than they would otherwise be [8]. Once again, an appropriate, conventional air pollution control strategy, this time for carbon monoxide, could address both local air pollution, climate change and, to a lesser degree, stratospheric ozone depletion.

A final argument that favors linking policies for global climate change and conventional air pollution is that a framework of laws currently exists to address the latter, while it may take some time to develop a similar body of law for climate change. There is already a major impetus to strengthen clean air standards, particularly in nonattainment areas of the United States such as southern California and in Arizona and Colorado. Citizens are far more likely to demand that their local air quality be improved immediately and galvanize their political leaders to do something about it than they are to mobilize to address a long-term issue such as climate change. Hence, it may be possible to begin slowing global warming sooner using this strategy than to wait for national greenhouse policy to be developed and international agreements to be completed.

It is clear from the foregoing discussion that energy policy plays a pivotal role in addressing the full range of climate and atmospheric problems. The most direct response to stabilizing carbon dioxide concentrations in the atmosphere is to improve the efficiency with which we utilize fossil fuels in the production and use of energy service. Yet the United States, while having made impressive gains in energy efficiency since 1973, still lags far behind western Europe and Japan. While the U.S. spends eleven percent of its GNP on energy, Japan and the Federal Republic of Germany spend closer to five percent of theirs [9]. Stated another way, in 1986, it required 20,700 kilojoules of energy for the United States to produce one dollar's worth of GNP, whereas it took Germany 11,300, and Japan only 9,800 kilojoules to create the same amount of economic activity [10]. As a result, the average German releases only 3.0 tons of carbon as carbon dioxide each year, while his American counterpart releases 4.9 tons [11]. Much of this difference can be accounted for by the fact that Germans live in smaller dwelling units that are more energy efficient, have better public transportation, drive more fuel-efficient cars, recycle more waste materials and utilize more cogeneration and district heating than do Americans. On the other hand, the unlimited speeds on the German Autobahn increase the amount of carbon dioxide and air pollutants.

Despite the efficiency gains of both Germany and the United States, there remain many opportunities for improving the efficiency of energy use further. In the residential and commercial sectors, new fluorescent lighting systems offer the same light output as traditional incandescent bulbs with only 20% of the electricity input. These new bulbs provide a soft, attractive glow, have a payback period of three years or less, and reduce the cooling loads in buildings substantially. Refrigerator designs which are now available in the U.S. provide the same amount of cooled space as the average American machine of five years ago with only 50-75% of the electricity input [12]. Recent efficiency laws enacted in the U.S. will improve the efficiency of many appliances, reducing the need for 20-40 large coal-fired power stations during the next thirty years. Some Japanese designs are even more efficient.

In the residential sector, new building designs are available which can reduce the heating loads per square meter for U.S. single-family dwellings by a factor of ten below the average of the current housing stock. In Sweden, where conservation of energy in buildings has long been considered important, new prefabricated buildings sold by Faluhus require only one-sixth the energy of the average Swedish housing stock [13].

Major improvements are also possible by modernizing industrial technology. The cement industry in the Federal Republic, for example, uses only two-thirds as much energy per ton of output as does the cement industry in the United Kingdom. The paper industry in the U.S. offers a similar advantage over many of its partners in the OECD. By contrast, the U.S. steel industry uses 50% more energy on average to make a ton of crude steel than does the industry in Japan [14]. Licensing, transfer and adoption of the most advanced, energy-efficient technology in each industrial sector offers the greatest potential for reducing global emissions and spurring economic growth.

The increased use of cogeneration for both electric power generation and district heating or process heat, which is now mandated in Germany, and is being used much more in the U.S., will significantly lower carbon dioxide emissions.

In the transportation sector, the most important opportunities for increased efficiency are in road and air transport. New jet engines and composite materials for aircraft construction offer the promise of advanced designs which can fly longer distances while burning half of the fuel per ton-km. Although current low fuel prices have discouraged the further development of such advanced airplanes, it is certainly feasible to introduce them over the next fifteen years. New car designs such as the VW Auto 2000, the Renault EVE and the Volvo LCP 2000 provide adequate proof that cars which use less than 4 liters of fuel per 100 km can deliver safe, attractive and efficient transportation for four or five passengers [15].

To summarize, not only would improved efficiency of fossil fuel use help slow the rate of climate change, reduce air pollution and acid rain, it would also improve the economic competitiveness of our economy and reduce the imported oil component of our rapidly growing trade deficit.

In a similar way, a comprehensive view of our energy and environmental issues would suggest that a strategy of fuel switching might be a useful way to simultaneously address several of these problems. While coal is much more abundant than natural gas in the U.S., there are circumstances for which substituting natural gas for coal would alleviate a number of problems. Natural gas contains no sulfur and produces no particulates when burned, a major advantage over coal for improving acid rain and local air quality. It also releases only about one-half the carbon dioxide per unit of heat that coal does, and is more efficient than coal in producing electricity. There appears to exist sufficient natural gas to act as a bridge over the next 40 years until we develop carbon-free energy technologies [16].

A number of promising alternative energy technologies are in development and are being implemented. The oldest harnessed renewable energy source, hydropower, supplied 21 percent of the world's electricity in 1986 [17]. Geothermal electric capacity now exceeds 5,000 megawatts worldwide, with an estimated 12,000 MW of additional geothermal heating power in use [18]. Nearly 600 megawatts of wind power is being added to global electric generating capacity annually, and India plans to have five gigawatts of wind capacity by the year 2000 [19]. Solar hot water heating provides a majority of Israel's needs, and is expanding worldwide. Solar thermal electric capacity in California stands at 194 MW at the end of 1988, with 80 more MW due to come on-line in 1989. The latter facility will be producing needed peaking power for air conditioning demand at a cost of 8 cents per kilowatt-hour. 300 MW of additional capacity are under contract [20]. The price and efficiency of photovoltaic generated electricity have shown dramatic improvement during the last few years. Several experimental crystalline cells have exceeded 30 percent efficiency [21], and costs of amorphous cells are dropping to $1.00 per peak watt [22]. Nuclear power is, after hydro, the second largest carbon-free electricity source. In 1986, 398 power plants with a capacity of 291 GW were in place, but the rate of growth has been slowing.

Unfortunately, our governmental structure for addressing this collection of environmental and economic problems is highly fragmented and poorly suited for seeing the interconnections among them. While most nations have an Environmental Ministry, there exist no Departments of Carbon Dioxide Control or Climate Protection. Domestic economic policy, trade issues, national security, energy policy, energy regulation and environmental regulation are viewed as distinct areas of governmental concern, and are accordingly administered by separate agencies. In the climate protection area alone, I have identified some 17 different existing federal agencies that might have some responsibility for global climate change alone [24]. (See Table 2.) As a result, many conflicting policies are simultaneously enacted, such as the rollback of auto fuel efficiency standards in 1988 at the same time that oil imports were increasing and concern over the greenhouse effect was growing [25].

Table 2

Global Climate Change

Proposed Agency-by-Agency Allocation of Responsibilities

Environmental Protection Agency

- Chair interagency committee and coordinate policy options studies
- Phase-out emissions of CFCs
- Identify other opportunities to reduce trace gases
- Study impacts of climate change due to sea level rise, and analyze the implications of warmer temperatures for human health, for air pollution, and for the natural environment
- Global tropospheric ozone policy
- Nitrous oxide (N2O) strategy
- Global analysis of emissions
- Integration with existing air pollution programs
- Examine the consequences of sea-level rise for salt water intrusion into coastal aquifers
- Study the consequences of sea-level rise on near coast landfills, toxic waste sites and sewage treatment plants

Department of Energy

- Implement tighter efficiency standards on appliances and buildings to promote energy efficiency
- Phase-out subsidies to publicly-owned utilities and energy companies to reflect more accurately the cost of energy to consumers
- Revise natural gas and wholesale electricity pricing to reflect more accurately marginal costs and promote efficiency (FERC)
- Revise energy supply priorities in research programs to emphasize renewable energy technologies and new efficiency technologies
- Conduct policy analysis on global energy use and alternatives
- Analyze implications of climate change for the energy system (e.g., change in patterns of electricity consumption and effect of sea level rise on energy facilities)
- Oversee energy efficiency goals for the federal government

State Department

- Advocate climate initiatives at UN agencies
- Support bilateral and multilateral agreements on research and action to address the greenhouse problem
- Consider climate change issues in U.S. development assistance programs
- Prepare for impact of climate change on refugees and increased aid
- Explore debt swapping and other approaches as mecha-nisms to preserve tropical forests

Treasury Department

- Support assessment of climate impacts in the lending process of the World Bank and other lending agencies supported by the U.S.
- Study carbon dioxide (CO2) tax

- Study gasoline tax and other consumption-based taxes

Interior Department

- Reform fossil fuel leasing policies
- Assess the implications of climate change for the management of national parks and other public lands
- Assess implications of rising sea-level on coastlines (USGS) and consequences for water projects (Bureau of Reclamation)

Corps of Engineers

- Determine consequence of sea-level rise on harbors, coastal waterways and beaches
- Evaluate implications of changed climate on water projects and navigable rivers

Agriculture Department

- Study implications of climate change for agriculture and the priorities of agriculture research
- Study implications of climate change for management of the national forests including increased fires
- Study agricultural practices as sources of increased methane and nitrous oxide emissions
- Assist in development of a global reforestation strategy
- Analyze multiple stresses on crops and forests

Transportation Department

- Improved standards on auto efficiency
- Assess impacts of climate change and sea-level rise on transportation system

Defense Department

- Assess implications of climate change for national security
- Reduce DOD energy use

Labor and Commerce Departments

- Examine shift in labor market in energy, agriculture and commerce as greenhouse strategies develop and implications for U.S. economy, employment and trade

Council of Environmental Quality

- Review greenhouse implications as part of the EIS process under NEPA

Coordinated science and policy research among:

Environmental Protection Agency

Department of Energy

U.S. Geological Survey

National Science Foundation

National Aeronautics and Space Administration

National Oceanic and Atmospheric Administration/National Climate Program

- Establish research goals for an improved understanding of global change and the earliest possible detection of a greenhouse fingerprint
- Develop long-range research programs to study and monitor global climate change over the next 30 years (Global Change 2020 Program)

The Congress is even further divided into committees and subcommittees whose jurisdictions do not even match those of the agencies and are inconsistent between the two houses. Even though there would be a greenhouse effect in every Congressional district, should it occur at all, Congress is a geographically representative body. As a result, it tends to focus on short-term regional interests rather than long-term global issues. This regionalism is reflected in the failure of the Congress to amend the Clean Air Act since 1977 because of differing consequences of acid rain provisions for various states. Despite this problem, it is encouraging that there are more than a dozen pieces of legislation introduced into the Congress that address everything from energy policy to population programs. Whether any of these proposals actually become law will depend in large measure on whether global issues are seen as a greater mutual threat than the local economic disruption their prevention will entail.

It remains to be seen to what extent the United States and other nations will act to reverse current trends in global and region-wide environmental deterioration. In this paper, I have tried to make the point that the most effective strategies will be those that simultaneously address several environmental and economic issues at once. Optimal solutions that utilize energy efficiency and alternative technologies are likely to prove to be the most cost-effective, but devising them is likely to prove to be a major challenge to our political and social institutions.

References

1. Ramanathan, V., Cicerone, R.J., Singh, H.B. and Kiehl, J.T., 1985. "Trace Gas Trends and Their Potential Role in Climate Change." *Journal of Geophysical Research* **90**: 5547-5566.

2. Seinfeld, J.H., 1989. "Urban Air Pollution: State of the Science." *Science* **243**: 745-752.

3. Schwartz, S.E., 1989. "Acid Deposition: Unraveling a Regional Phenomenon." *Science* **243**: 753-763.

4. Hansen, J., Fung, I., Lacis, A., Lebedeff, S., Rind, D., Ruedy, R. and Russell, G., 1987. "Prediction of Near-Term Climate Evolution: What Can We Tell Decision Makers Now?" *Preparing for Climate Change, Proceedings of the First North American Conference on Climate Change: A Cooperative Approach*. The Climate Institute, Washington, D.C., pp. 35-47.

5. Seinfeld, J.H., Op. Cit.

6. Ibid.

7. Cicerone, R.J. and Oremland, R.S., 1988. "Biogeochemical Aspects of Atmospheric Methane." *Global Biogeochemical Cycles* **2**(4): 299-327.

8. Ibid.

9. World Resources Institute, International Institute for Environment and Development and United Nations Environment Program, 1988. *World Resources 1988-89*. Basic Books, Inc., New York.

10. Ibid.

11. Ibid.

12. Goldemberg, J., Johanssen, T., Williams R. and Reddy, A., 1987. *Energy for a Sustainable World*. World Resources Institute, Washington, D.C.

13. Ibid.

14. Ibid.

15. Bleviss, D.L., 1988. *The New Oil Crisis and Fuel Economy Technologies: Preparing the Light Transportation Industry for the Nineteen Nineties*. Quorom Books, Greenwood.

16. Burnett, W.M. and Ban, S.D., 1989. "Changing Prospects for Natural Gas in the United States." *Science* **244**: 305-310.

17. World Resources Institute, 1988. Op. Cit.

18. Ibid.

19. Ibid.

20. Luz International, Ltd., 1989. *Solar Electric Generating System (SEGS), Information Overview.*

21. Hubbard, H.M., 1989. "Photovoltaics Today and Tomorrow." *Science* **244**: 297-304.

22. Ogden, J. and Williams, R., 1989. *Solar Hydrogen: Moving Beyond Fossil Fuels*. World Resources Institute, Washington, D.C.

23. World Resources Institute, 1988. Op. Cit.

24. Moomaw, W.R., 1989. "Near-Term Congressional Options for Responding to Global Climate Change," *The Challenge of Global Warming*, Abrahamson, D.E. (ed.). Island Press, Washington, D.C., pp. 305-326.

25. Schneider, S.H., 1989. "The Greenhouse Effect: Science and Policy." *Science* **244**: 771-781.

THE ROLE OF FORESTS IN ADDRESSING THE CO_2 GREENHOUSE

Gregg Marland

Environmental Science Division
Oak Ridge National Laboratory
P.O. Box 2008
Oak Ridge, TN 37831-6038

If we should conclude that increasing atmospheric concentrations of greenhouse gases, principally carbon dioxide, truly threaten adverse changes in the global climate, there are four fundamentally different ways in which we might choose to respond. Obviously the kind and scale (and cost) of the responses we are willing to consider implementing will depend on our perception of urgency—how serious we think the problem is and how much time we think we have to respond. Although the drought and high temperatures of the past summer have sensitized society to the greenhouse issue, studies to date have not turned up any easy solutions. Carbon dioxide is an essential and high-volume by-product of fossil-fuel burning and it is recognized that, so long as we rely on fossil fuels to provide useful energy, we will be discharging CO_2 to the atmosphere.

I will characterize the possibilities for responding to the "CO_2 problem" under the headings: fix, change, tinker, and cope.

By "change" I mean switching to an energy system that discharges smaller quantities (preferably none) of infrared-absorbing gases to the atmosphere. This includes solar and nuclear energy systems along with the more efficient use of energy and substitution of natural gas for coal.

By "cope" I mean simply letting the climate evolve and trying to accommodate the changes as they occur. This has also been characterized as the "procrastinate" option but it is possible that we will make an affirmative decision that this is the most reasonable approach to select. Of course the longer we delay a decision, the larger the role that coping will play in our ultimate menu of responses! This option could eventually involve immense environmental modifications and land-use changes as we try to supply water to drier agricultural areas while keeping it out of the streets of Galveston. It might involve abandoning some agricultural lands and establishing new ones in areas now restricted by temperature, rainfall, or length of growing season. It might involve the relocation of large populations such as those in Bangladesh and of smaller populations such as those on South Pacific atolls. I do not intend to dwell on the concept of adaptation but want to ensure that we have some vision of the scale of alternatives when I

Published by Elsevier Science Publishing Company, Inc.
Global Climate Change Linkages
James C. White, Editor

describe the possibility of massive reforestation. We do need to keep in mind that there could be widely different impacts in different regions across this political globe.

The "tinker" options are those that would intervene in the climate system to try to balance the changes wrought by an increase in greenhouse gases. The most obvious possibility is to increase the global albedo. This is the approach being taken to a smaller scale greenhouse problem when a light colored piece of folded cardboard is placed behind a car windshield to limit heating on a sunny day. Climate model studies of nuclear-winter scenarios and of the albedo feedback observed as a consequence of melting ice and snow suggest that it is theoretically possible on a global scale as well. According to Ramanathan (1988), "An increase in the planetary albedo of just 0.5% is sufficient to halve the greenhouse effect of CO_2 doubling." Because less than half of the incoming solar radiation ever reaches the earth's surface, this albedo change would be most practically done as high as possible in the atmosphere. It has been shown, for example, that stratospheric aerosols can have direct radiation effects on climate and Bryson (1988) argues that the intensity of volcanic aerosol emissions has been a major factor in historic climatic variability (see also Harvey, 1988, and Self and Rampino, 1988). The maximum surface cooling due to the El Chichon volcanic eruption in Mexico on April 4, 1982 has been estimated at 0.3° – 0.4°C (Ramanathan et al., 1987).

There are hazards in trying to predict what might result from large-scale efforts to alter the earth's albedo. Henderson-Sellers et al. (1988) used a climate model to study the climatic effect of a major land-use change. In an experiment that has further application to the discussion that follows, they modeled the effect of complete deforestation of the Amazon; 5×10^6 km^2 of tropical moist forest was replaced with grass, with the attendant increase in albedo, plus changes in surface roughness and moisture holding capacity of the soil. In this study, changes in surface albedo, evapotranspiration, and cloud cover interacted to produce no discernable change in local or global-scale temperature, planetary albedo, or planetary precipitation. Henderson-Sellers et al. (1988) described, also, a similar experiment with a different climate model which yielded a 2°C increase in surface air temperature. The possibilities for unanticipated impacts are staggering.

"Fixes" are strategies that would allow us to continue burning fossil fuels while assuring that the carbon that is discharged as CO_2 ends up somewhere other than in the atmosphere. The challenges are how to collect the carbon and what to do with it once collected. We have industrial experience that demonstrates we can collect CO_2 from a fossil-fuel-burning exhaust stream and we can move it large distances by pipeline. The literature contains a variety of creative suggestions for what might be done with the CO_2 once collected, but insertion into the deep ocean is the most promising idea to date. Albanese and Steinberg (1980) have shown that mechanical/chemical collection from the dilute CO_2 mixture of ambient air is a losing enterprise and that any such collection scheme must focus on concentrated streams such as those at power-plant stacks. All of these fixes are very expensive and encounter serious problems for carbon disposal.

All of this having been said by way of preface, the "fix" I want to focus on here is the possibility of taking advantage of the carbon-fixing talents of green plants and collecting

CO_2 from the atmosphere as carbon in biomass – that is, as trees. Is it feasible to stimulate the growth of forests to take up in woody matter enough carbon to offset the discharge of CO_2 from fossil-fuel burning?

It is useful, in beginning, to think about tree-growing on two scales. We can consider, on a global scale, the magnitude of the effort required to offset all CO_2 emissions from fossil-fuel burning. We can also consider, on a per-capita or per-power-plant scale, the effort required to offset some increment of those CO_2 emissions. Let me make clear at the outset, as Dyson and I did when we first explored this topic in the late 1970's (Dyson, 1977; Dyson and Marland, 1979), that it is not possible to continually plant more trees to take up CO_2 from more fossil-fuel burning. To take up carbon in growing biomass is a transient means to halt or reverse the atmospheric CO_2 increase while we switch to an energy system that does not discharge infrared-active gases, or perhaps to slow the growth rate of atmospheric CO_2 while we figure out what needs to be done over time scales longer than the life-cycle of trees (see also Marland, 1988).

We should know first how much carbon dioxide we have to deal with. During 1986 global total emissions from fossil-fuel burning and cement manufacture amounted to just over 5.5 billion metric tons (tons throughout this paper always means metric tons) of carbon (Figure 1). Our experience over the last 30 years is that the cumulative increase in atmospheric CO_2 has been equivalent to about 55% (Emanuel et al., 1985; Bolin, 1986) of the carbon so discharged (compare Figure 2. with Figure 1). In trying to explain why this is so, Firor (1988), among others, has suggested that if we could reduce these fossil-fuel-based emissions by roughly half we "would come close to stabilizing the atmospheric burden of CO_2." A detailed study with a carbon cycle model would be required to learn exactly how atmospheric concentrations would respond to halving the fossil-fuel source term, although it is clear that in 1960, when this source term was half of the current value, the atmospheric increase in CO_2 (as measured at Mauna Loa observatory) still represented roughly 55% of the carbon emitted. The qualitative expectation is that halving the rate of carbon emissions would result in temporary relief in the growth of atmospheric CO_2 but that this relief would be short-lived and we would soon have to reduce emissions further in order to keep the atmospheric concentration from again increasing. Without arguing details, then, and recognizing that there are alternative views, we will assume that, if an individual, nation, or global society wanted to eliminate their contribution to increasing atmospheric CO_2 concentrations, they should seek to eliminate or offset their full contribution, not just the 55% that is currently being observed to accumulate in the atmosphere. This means offsetting 5.5×10^9 tons of carbon per year, for a global total, or 5.0 tons per person per year for a resident of the U. S. (4.1 tons per person per year in Canada). With net global primary productivity estimated at 50 to 60×10^9 tons C per year (SCEP, 1970), we are looking for a 10% increase in net carbon assimilation by the biosphere.

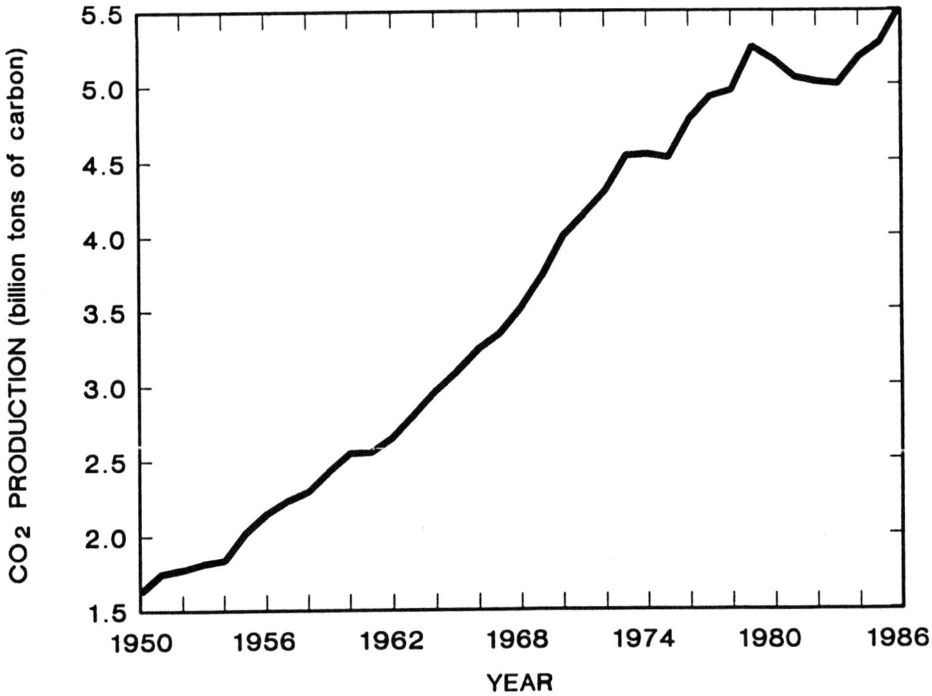

Figure 1: Global carbon dioxide emissions from fossil fuel burning and cement manufacture, 1950 to 1986. (in billions of metric tons of carbon)

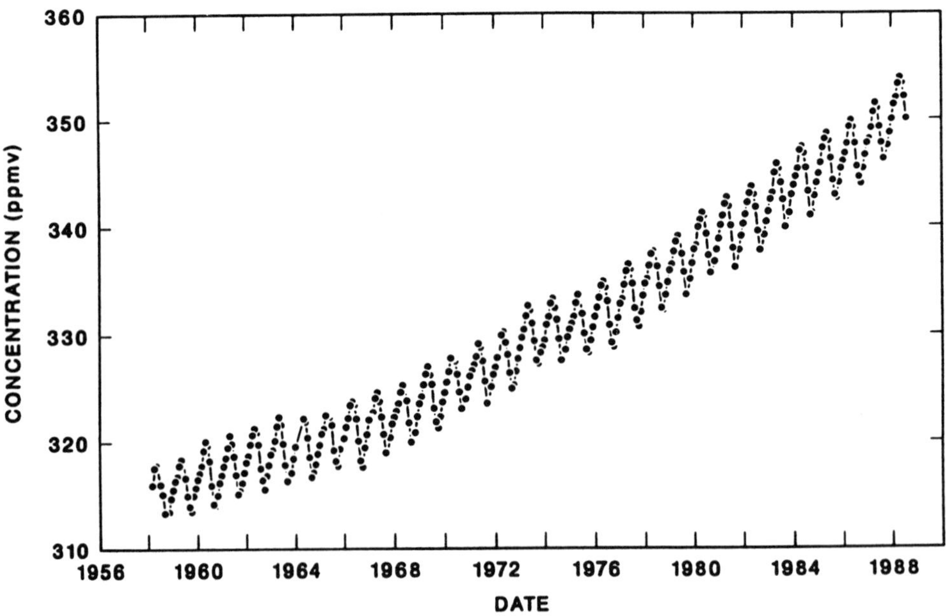

Figure 2; Monthly average values for atmospheric carbon dioxide concentration at Mauna Loa Observation. (Data from C. D. Keeling)

Table 1 provides focus for the problem. The table reveals that with nine countries we account for over half of the world's land surface and population and roughly two-thirds of its fossil-fuel-related CO_2 emissions and existing closed forest. It emphasizes the great disparity in the per-capita rate of CO_2 emissions and shows that, whereas countries like Zaire and Brazil have very large areas of existing forest in proportion to their industrial CO_2 emissions, countries like West Germany and Japan would be much more taxed to consider a forest-based solution to CO_2 emissions.

The problem from a CO_2 perspective is not only to envision where reforestation might occur but, more immediately, to arrest deforestation that is continuing now. Our best estimate (Houghton et al., 1987) is that land use changes are currently resulting in the discharge of 1.0 - 2.6 x 10^9 tons of carbon per year to the atmosphere (mostly from deforestation in the tropics) – 20% or more of the quantity from fossil-fuel burning. Estimates of the 1980 rate of destruction of closed, primary tropical forest range from 4.5 to 11.9 x 10^6 hectares per year (Detwiler and Hall, 1988) with some recent literature suggesting that the current value may be even higher (e.g. Malingreau and Tucker, 1988). Figure 3 shows that the developed countries have cleared extensive areas of forest, a process which is only now progressing in many developing countries. Figure 4 illustrates that in south and southeast Asia, for example, an expanding population is resulting in clearing of forest to provide agricultural land – a process that would be difficult to reverse to a significant degree. Clearing of forests throughout the tropics is progressing for a variety of reasons, but most are ultimately related to population pressures. Nonetheless, forest area in the United States and Europe is apparently increasing now as improvements in agricultural productivity lead to abandonment of marginal agricultural land.

Matthews (1983) has provided an estimate of the total area of forest cleared to date (701 x 10^6 ha) – a value which lets us compare reforestation requirements with the extent of land which at one time supported forest but is now used for some other purpose (Table 2). Estimates of total forest clearing from Williams (1988) do not exactly match those of Matthews but are remarkably close and show how this area is distributed geographically (Table 3).

With this perspective of the area of forest and forest change globally, we inquire into the rate at which forests can take up carbon. Data from the U. S. Forest Service (1982) show that in 1977 the average net annual growth rate of U. S. commercial forest was 3.15 m^3 ha^{-1} yr^{-1}, which translates to a carbon uptake of approximately 0.8 tons carbon ha^{-1} yr^{-1}. This is carbon storage in stemwood and is thus a low estimate of total carbon storage in forests (by perhaps a factor of 2), but it is a useful number if our interest is in long-term storage of carbon in woody biomass. It suggests that in the U. S. it would require 6.35 hectares per capita to offset the current fossil-fuel-related CO_2 emissions in the wood of forests comparable to current average commercial forest. With a population of 246 million, it would thus require 1560 million hectares of these forests to offset current U. S. emissions. This is 70% larger than the land area of the U. S.

Table 1

AN 11-COUNTRY SAMPLE TO FOCUS THE CO_2 PROBLEM

	Land Area (10^6 ha)	Closed Forest (10^6 ha)	1989 Population (10^6)	1986 CO_2 Emissions (10^6 tons C)	fossil fuel/ area of CO_2 emissions/ closed forest (tons Cyr^{-1}/ha forest)	1986 per capita emissions (tons C/person)
1. U.S.S.R	2227	792	289	1011	1.3	3.6
2. U.S.A	917	195	246	1202	6.2	5.0
3. China	933	98	1111	554	4.4	0.53
4. India	297	52	813	144	2.0	0.18
5. Mexico	192	46	87	74	1.5	0.91
6. Brazil	846	357	147	53	0.1	0.38
7. Canada	922	264	27	105	0.4	4.1
8. Indonesia	181	114	179	28	0.2	0.17
9. Japan	38	24	123	256	10.7	2.1
Sum (1-9)	6557	1942	3022	3427		
% of Total	50.10	67.20	58.54	62.18		
10. Zaire	227	106	34	1	0.01	0.03
11. W. Germany	24	7	60	186	26.6	3.1

Area and population data from World Resources Institute, 1988[b].

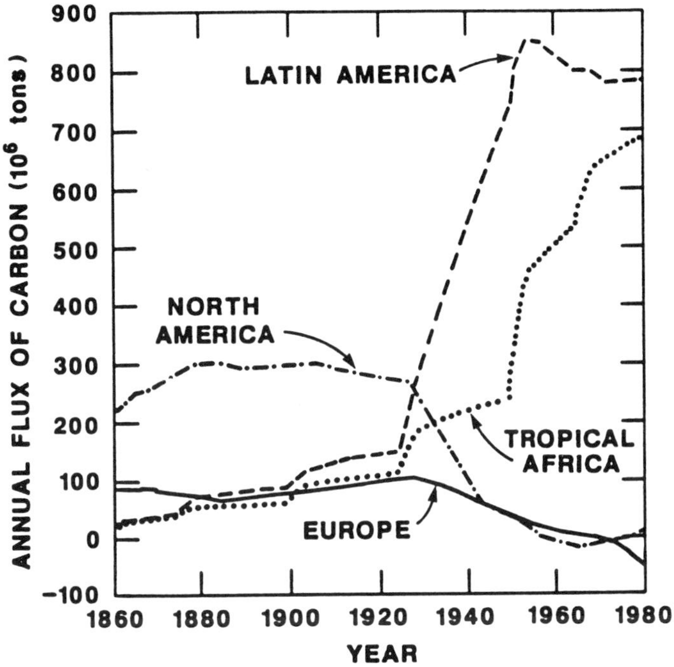

Figure 3: Net flux of carbon to the atmosphere as a result of land use changes and harvest. From Houghton et al., 1983.

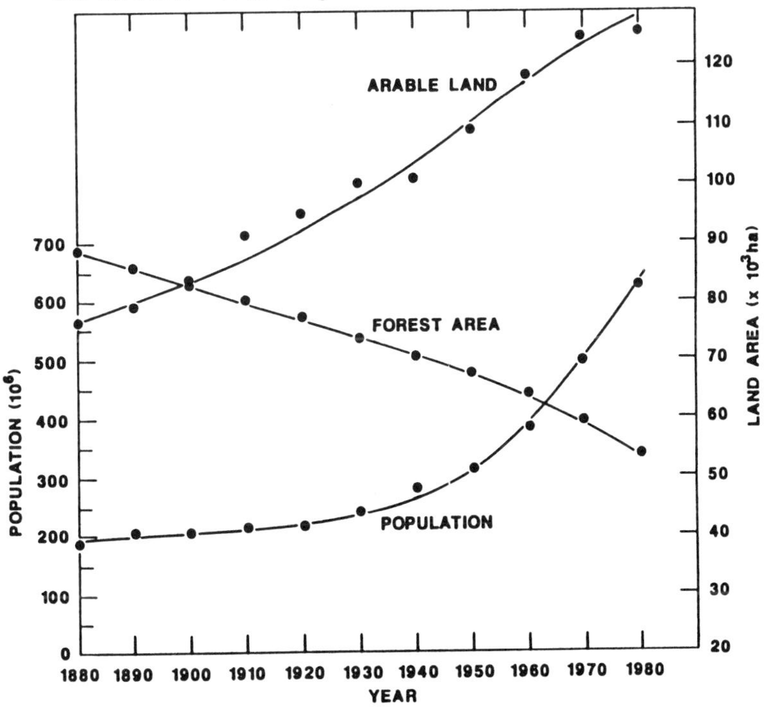

Figure 4: Changing land use and population since 1880 in Pakistan, Bangladesh, Burma, Malaysia, Brunei, and Northern India. Data from Richards et al. 1987.

Table 2

Comparison of Areal Estimates of Major Ecosystems. From Matthews, 1983.

	Pre-agricultural Vegetation Area (10^6 ha)	Present Land Cover Area (10^6 ha)	Ecosystem Reduction Area (10^6 ha)
Total forest	4628	3927	701
Tropical rain forest	1277	1229	048
Other forest	3351	2698	653
Woodland	1523	1310	213
Shrubland	1299	1212	087
Grassland	3390	2743	647
Tundra	734	734	000
Desert	1582	1557	025
Cultivation	093*	1756	-1663

Table 3

Estimated Clearing, Through 1978, of Forest Plus Woodland
(High estimate from Williams, 1988)

Region	Area (10^6 ha)
North America	110.7
Central America	28.8
Latin America	92.5
Oceania	38.0
USSR	109.5
Europe	49.7
Asia	300.6
Africa	75.9
Total	805.7

Table 4

PRODUCTIVITY INCREASES ATTRIBUTABLE TO INTENSIVE PLANTATION MANAGEMENT

Case	Maximum Mean Annual Yield (Tons Carbon ha^{-1} yr^{-1})
Douglas Fir in Washington	
Natural stands	2.8
Silvicultural treatments	
Plantation establishment	3.6
Nitrogen fertilization	4.4
First-generation genetics	4.9
Target	12.5
Loblolly Pine in North Carolina Pocosins	
Natural stands	1.8
Silvicultural treatments	
Drain and plant	3.5
Bedding	4.3
Preplant phosphorus	5.3
Nitrogen fertilization	5.9
First- and second-generation genetics	7.2
Target	15.0

BASED ON FARNUM ET AL. (1983)

The growth rate of trees is, however, a function of site and stand attributes and considerably higher growth rates can be achieved in managed plantations. The Forest Service (1982) estimated that 1977 productivity could have been increased by two-thirds simply by assuring that the forests were fully stocked. By using various combinations of species selection, genetic screening, spacing control, thinning, pruning, weed control, fire and pest control, fertilization, irrigation, harvest cycle-time control, and site preparation, growth rates can be vastly enhanced (see e.g. Table 4). Carbon uptake as high as 14 tons carbon ha^{-1} yr^{-1} have been achieved for hybrid cottonwoods in western Washington (Heilman and Stettler, 1985).

As a model for further illustration, let us consider American sycamores grown on a tree plantation in Georgia (Steinbeck and Brown, 1976). Planted on 1.3 m centers and harvested on a 4-year rotation, these sycamores regrow from the stumps and can average 7.5 tons carbon ha^{-1} yr^{-1} as total biomass. The value is a reasonable target for a variety of fast-growing species on a wide variety of sites, with intensive management. At 7.5 tons carbon $ha^{-1} yr^{-1}$, the requirement for offsetting CO_2 emissions in the U. S. would be only 0.67 ha per person, or 164×10^6 ha for the nation. This is less than the current forested area of the U. S. and only 48% more than the total of North American forest clearing to date. These numbers make it begin to appear possible that reforestation could indeed offset some important fraction of fossil-fuel CO_2 emissions if we focus on young, vigorous trees in managed plantations. We might even envision achieving higher carbon uptake per hectare through genetic manipulation of characteristics such as growth rate, nutrient-use efficiency, drought tolerance and pest resistance.

To offset the global total fossil-fuel emissions of 5.5 billion tons carbon with trees growing at the rate of American sycamore in Georgia would require 733 million hectares, an area only slightly smaller than the continent of Australia. (An estimated additional 5-8% would actually have to be allocated to cover the energy costs of plantation establishment and operation (J. Ranney, personal communication, 1989).) On a more manageable scale, to offset the CO_2 emissions from a 1000-MW coal-fired generating plant operating at 38% thermal efficiency and 70% availability would require 190 hectares of trees per installed MW, or a plantation of sycamore equivalents to a radius of 24.7 km surrounding the power plant. (This does include the energy cost of coal mining and transportation.)

It is important to emphasize that the high growth rates mentioned here are for young vigorous plantations. For forests there is an explicit trade-off between high rates of carbon uptake and high rates of carbon storage. Mature forests have a large mass of carbon storage but very small rates of net carbon assimilation and Cooper (1983) has estimated that a forest managed for maximum sustained yield will contain only one-third the biomass of a mature stand. The consequence is that to maintain high rates of carbon uptake, the American sycamore plantation was harvested every 4 years. To maintain the carbon sink it would be necessary to either produce trees that could maintain high rates of growth for longer times or to harvest the trees and prevent their oxidation back to CO_2 while another crop was growing.

In fact, a small U. S. power company (AES Thames) has recently announced a reforestation project for a CO_2 offset. To offset 15 million tons of carbon to be discharged over the 40-year life of a 180-MW coal-fired power plant planned for Uncasville, Connecticut, they have contracted with CARE for a reforestation project in Guatemala. The project will involve 40,000 small-holder farmers, 52 million trees over 100,000 hectares, and a variety of objectives, including woodlots and soil conservation, in addition to the CO_2 offset. With U.S. $2 million from the power company and U.S. $14.3 million in cash or in kind from other project participants (including CARE, the Guatemalan government, the U.S. Agency for International Development, and the Peace Corps), the anticipated cost of the project is less than 6% of the capital cost of the power plant (World Resources Institute, 1988a). It would be much more were it not for the free labor of the local landowners.

As a further possibility, we might reconsider storage or disposal of the trees once grown. Since wood-burning takes carbon from the air during tree growth and returns it during wood burning, there is no net CO_2 release (aside from the fuel consumed during forest management and harvest). If the harvested wood could be fully substituted for fossil fuels there would be no net CO_2 emissions to the atmosphere as long as the trees were grown on a sustained-yield basis.

Having considered the possibility of planting additional trees to take up atmospheric CO_2, could we then burn those trees (instead of additional fossil fuel) to maintain the current atmospheric CO_2 level? One major obstacle is the low conversion efficiency that is characteristic of wood-burning facilities. Durst (1986) reports that for wood-fired power plants in the Philippines each MW of capacity typically requires 400 hectares of intensively managed, high-productivity plantations operated on sustained-yield basis. Frankena (1987) suggests that the thermal efficiency of conversion is 25% at best. The low efficiency is largely a consequence of high fuel moisture content and Ostlie (1988) has recently described a scheme for burning trees whole, using waste heat for pre-combustion drying, for which he reports 34% conversion efficiency. Ostlie's plant would appear to reduce plantation fuel-wood requirements to 217 ha/MW for trees growing at the rate of American sycamores in Georgia. This discussion of a fix scenario has shifted to one that bears elements of both fix and change and therein gains utility over a longer time period.

One important point for reforestation prospects is that the fast-growing, short-rotation plantations do not resemble mature forest but are visually more like an agricultural crop and functionally a hybrid of the two systems. Although they could provide many benefits, besides wood fuels and CO_2 sinks, we should not be misled to expect the emotional and recreational benefits of a mature forest. Likewise they do not have the carbon storage capacity of a mature forest and it would be counterproductive, as a CO_2 solution, to harvest mature forests and to replace them with young fast-growing plantations. What is needed is additional forest to either remove carbon from the atmosphere and/or to provide a sustainable substitute for fossil fuels. Also, high growth rates require high-quality sites and it is high-quality sites that will respond best to silvicultural inducements. Tree plantations will be more expensive and less effective on marginal land.

Finally, I acknowledge that I have not addressed the linkages that are the major theme of these proceedings. This discussion of the prospects for increasing the yield of forests occurs in a setting where concern is being voiced over declining forest productivity because of acid precipitation, ozone, and organic and trace metal contaminants. We cannot yet know how possible changes in temperature and rainfall will impact forests and there are ongoing but inconclusive studies of the direct effects of increasing atmospheric CO_2 on tree growth. Major changes in forest area could in turn impact albedo and water budgets. Changing land-use patterns and fertilizer usage may impact emissions of nitrous oxide, methane, and other greenhouse gases. The linkages are indeed complex and need further consideration.

To conclude, I would simply reiterate that the magnitude of current CO_2 emissions is immense and any program that will offset or significantly affect net emissions will likewise be immense in both scale and cost. We are not likely to replace all of the forests ever cleared with new forests growing at the rate of American sycamores on plantations in Georgia. Yet it is physically possible to have a significant effect through forest policy from both increasing the mass of carbon stored in biomass and increasing substitution of wood fuels for fossil fuels. And, if climate changes are to be as dramatic as many believe, we must ask what the other choices are for either dealing with atmospheric CO_2 or coping with the attendant changes in climate.

On the other hand, if the question is whether we can continue, over the long term, to burn fossil fuels and at the same time avoid increasing atmospheric CO_2, the realistic answer with respect to forestry-related solutions is "no".

Acknowledgment

The inquiries summarized here were supported by the Carbon Dioxide Research Division, Office of Basic Energy Sciences, U.S. Department of Energy, under contract DE-AC05-840R21400 with Martin Marietta Energy Systems, Inc.

References

Albanese, A. S. and M. Steinberg. 1980. *Environmental Control Technology for Atmospheric Carbon Dioxide*, Final Report, U.S. Department of Energy, DOE/EV-0079, 51 pp.

Bolin, B. 1986. "How Much CO_2 Will Remain in the Atmosphere?" In: B. Bolin, B. R. Doos, J. Jager, and R. A. Warrick (eds.), *The Greenhouse Effect, Climatic Change, and Ecosystems*, pp. 93-156, J. Wiley and Sons, NY.

Bryson, R. A. 1988. "What the Climatic Past Tells Us About the Environmental Future," In: National Geographic Society, *Earth '88: Changing Geographic Perspectives*, National Geographic Society, Washington, D.C., pp. 230-247.

Cooper, C. F. 1983. "Carbon Storage in Managed Forests," *Canadian Journal of Forest Research* v. 13, pp. 155-166.

Detwiler, R. P., and C. A. S. Hall. 1988. "Tropical Forests and the Global Carbon Cycle", *Science* v. 239, pp. 42-47. See also the letters to *Science* v. 241, pp. 1736-1739 (Sept. 30, 1988).

Durst, P. B. 1986. "Wood-Fired Power Plants in the Philippines: Financial and Economic Assessment of Wood-Supply Strategies," *Biomass* v. 11, pp. 115-133.

Dyson, F. J. 1977. "Can We Control the Carbon Dioxide in the Atmosphere?," *Energy* v. 2, pp. 287-291.

Dyson, F. J. and G. Marland. 1979. "Technical Fixes for the Climatic Effects of CO_2," In: W. P. Elliott and L. Machta (eds.), *Workshop on the Global Effects of Carbon Dioxide from Fossil Fuels*, Miami Beach, Florida, March 7-11, 1977, U.S. Department of Energy, CONF-770385, pp. 111-118.

Emanuel, W. R., I. Y.-S. Fung, G. G. Killough, B. Moore III, and T.-H. Peng. 1985. "Modeling the Global Carbon Cycle and Changes in the Atmospheric Carbon Dioxide Levels," In: J. R. Trabalka (ed.), *Atmospheric Carbon Dioxide and the Global Carbon Cycle*, DOE/ER-0239, pp. 141-173, U. S. Department of Energy.

Farnum, P., R. Timmis, and J. L. Kulp. 1983. "Biotechnology of Forest Yield," *Science* v. 219, pp. 694-702.

Firor, J. 1988. "Public Policy and the Airborne Fraction," *Climatic Change* v. 12, pp. 103-105.

Frankena, F. 1987. "Rethinking the Scale of Biomass Energy Conversion Facilities: The Case of Wood-Electric Power," *Biomass* v. 14, pp. 149-171.

Harvey, L. D. D. 1988. "Climatic Impact of Ice-age Aerosols," *Nature* v. 334, pp. 333-335.

Heilman, P. E. and R. F. Stettler. 1985. "Genetic Variation and Productivity of Populus Trichocarpa T and G and its Hybrids. II. Biomass Production in a 4-year Plantation," *Canadian Journal of Forest Research* v. 15, pp. 384-388.

Henderson-Sellers, A., R. E. Dickinson, and M. F. Wilson. 1988. "Tropical Deforestation: Important Processes for Climate Models," *Climatic Change* v. 13, pp. 43-67.

Houghton, R. A., J. E. Hobbie, J. M. Melillo, B. Moore, B. J. Patterson, G. R. Shaver, and G. M. Woodwell. 1983. "Changes in the Carbon Content of Terrestrial Biota and Soils Between 1860 and 1980: A Net Release of CO_2 to the Atmosphere," *Ecological Monographs* v. 53, pp. 235-262.

Houghton, R. A., R. D. Boone, J. R. Fruci, J. E. Hobbie, J. M. Melillo, C. A. Palm, B. J. Peterson, G. R. Shaver, G. M. Woodwell, B. Moore, D. L. Skole, and N. Myers. 1987. "The Flux of Carbon From Terrestrial Ecosystems to the Atmosphere in 1980 Due to Changes in Land Uses: Geographic Distribution of the Global Flux," *Tellus* v. 39b, pp. 122-129.

Malingreau, J. P. and C. J. Tucker. 1988. "Large-Scale Deforestation in the Southeastern Amazon Basin of Brazil," *Ambio* v. 17, pp. 49-55.

Marland, G. 1988. *The Prospect of Solving the CO_2 Problem through Global Reforestation*, TR039, U.S. Department of Energy, Office of Energy Research, Carbon Dioxide Research Division, 66 pp.

Matthews, E. 1983. "Global Vegetation and Land Use: New High-Resolution Data Base for Climate Studies," *Journal of Climate and Applied Meteorology* v. 22, pp. 474-487.

Ostlie, L. D. 1988. *The Whole Tree Burner: A New Technology in Power Generation*, Energy Performance Systems, Inc., Minneapolis, Minn.

Ramanathan, V. 1988. "The Greenhouse Theory of Climate Change: A Test by an Inadvertent Global Experiment," *Science* v. 240, pp. 293-299.

Ramanathan, V., L. Callis, R. Cess, J. Hansen, I. Isaksen, W. Kuhn, A. Lacis, F. Luther, J. Mahlman, R. Reck, and M. Schlesinger. 1987. "Climate-Chemical Interactions and Effects of Changing Atmospheric Trace Gases," *Reviews of Geophysics* v. 25, pp. 1441-1482.

Richards, J. F., E. S. Haynes, J. R. Hagen, E. P. Flint, J. Arlinghaus, J. B. Dillon, and A. L. Reber. 1987. *Changing Land Use in Pakistan, Northern India, Bangladesh, Burma, Malaysia, and Brunei, 1880-1980*, U.S. Department of Energy Report (in press).

SCEP. 1970. *Man's Impact on the Global Environment: Assessment and Recommendations for Action, Report of the Study of Critical Environmental Problems*, Cambridge, MA, MIT Press.

Self, S. and M. R. Rampino. 1988. "The Relationship Between Volcanic Eruptions and Climate Change: Still a Conundrum?" *EOS* Feb. 9, 1988, pp. 74-86.

Steinbeck, K. and C. L. Brown. 1976. "Yield and Utilization of Hardwood Fiber Grown on Short Rotations," In: *Applied Polymer Symposium* No. 28, New York, J. Wiley and Sons, pp. 393-401.

Turner, B., R. Kates, and G. White (eds.). *The Earth as Transformed by Human Action*, Cambridge University Press (in press).

U. S. Forest Service, Department of Agriculture. 1982. *An Analysis of the Timber Situation in the United States, 1952-2030*, Forest Resource Report No. 23, December.

Williams, M. 1989. "Global Deforestation," Personal Communication.

World Resources Institute. 1988a. *Power Company to Fund Reforestation to Offset Carbon Dioxide Emissions, Slow Greenhouse Effect*, press release, World Resources Institute, Washington, D.C., Oct. 11, 1988.

World Resources Institute and the International Institute for Environment and Development. 1988b. *World Resources 1988-89*, J. A. Brewster, project director, Basic Books, Inc., NY.

ON THE BENEFITS OF SIMULTANEOUS MEASUREMENTS – AN EXAMPLE

Lester Machta

National Oceanic and Atmospheric Administration
8060 13th Street
Silver Springs, MD 20910

Abstract

The benefits of simultaneous monitoring of air qualities is illustrated by measurements of carbon dioxide and methane. The interpretation of the resulting data in the Arctic spring indicates a several-fold larger release to the air of methane from combustion associated activities than has appeared in the literature.

The importance of examining the chemical and physical interactions of pollutants, the joint consequences of many environmental insults caused by air pollution, and the feedbacks from control measures has a counterpart in monitoring. Here, simultaneous measurements of many substances lead to interpretations which would be difficult if each substance were studied separately. It goes beyond simply making as many measurements as possible at the same place; there should, to the extent possible, also be a concerted effort to sample over the same interval with instruments having similar response times. The benefit of simultaneous measurements is illustrated below; a new and provocative estimate of anthropogenic emissions of methane is deduced from concurrent sampling with CO_2.

In recent years the evidence for increasing atmospheric concentrations of methane with time has become unequivocal [1,2,3,4,5]. This increase is of consequence because methane is a greenhouse gas. The sources of methane and of its secular increase are of concern if there is any consideration of imposing controls to avoid the increase.

Conventional wisdom attributes most of the atmospheric methane, as well as most of the annual increase, to biogenic sources, although some of the biogenic emissions (e.g., from rice paddies) may be increasing with time due to anthropogenic stimulation. Less than 10% of the atmospheric methane is thought to derive from fossil fuels, either by leakage of natural gas between mining and combustion or from combustion itself. It is suggested below that the fossil sources of methane in the air may be many times greater than heretofore generally proposed.

Copyright 1989 by Elsevier Science Publishing Company, Inc.
Global Climate Change Linkages
James C. White, Editor

In the March-April period, simultaneous measurements of carbon dioxide and methane were made both aboard aircraft under the AGASP-II (Arctic Gas and Aerosol Sampling Program) and at ground level at the NOAA background monitoring station at Point Barrow, Alaska [6,7]. This period is favorable for investigating the relative sources of these gases since minimal local biogenic sources and little chemical transformation and uptake are likely with snow and ice covering the Arctic Basin and adjoining land area. The aircraft generally flew over the Arctic ice not far from Point Barrow and Alert, Canada at altitudes up to about eight kilometers.

Figure 1 [6] presents a plot of the simultaneous concentrations found by the aircraft as both gases varied in concentration with altitude. Part of the high (0.97) correlation is due to the decrease with altitude of both gases and part due to excursions of high values in haze layers. Figure 2 [7] represents the relationship found for two selected periods during which simultaneous measurements were made at ground level at Point Barrow. The selection involved picking a period during which the wind direction did not blow from the nearby village of Point Barrow towards the observatory and wind speeds exceeded 5 ms^{-1}. Again the correlation coefficients are very high, 0.93 and 0.94.

The high correlations after lengthy transport suggest that both gases originate in the same area. In the calculations below, it is assumed that both are connected with fuels for combustion or combustion itself and are typical of the entire world. Later the validity of such assumptions will be treated. The aim of the calculations is a determination of the global methane emission from combustion for comparison with previous estimates.

Rotty (1988) estimates the global emission of carbon dioxide from combustion of fossil fuels – coal, oil, natural gas or methane, and peat – to be about 5.4 gigatons of carbon (1 gigaton = 10^{15} grams) in 1986. A similar number for the emission of methane to the air from direct anthropogenic sources was estimated to be about 0.03 gigatons of carbon in methane [2]. The correlations shown in Figure 1 yield a ratio of 0.018 ppm methane for each 1 ppm of CO_2 in Figure 2 and 0.018 and 0.019 ppm methane for each 1 ppm carbon dioxide, respectively. An average value of 0.018 ppm is used below. A ratio of 0.018 ppm methane to 1 ppm carbon dioxide is the same as a ratio of carbon in each gas of 0.044 g in methane for each 1 g in carbon dioxide.

The global emission of methane from the same sources as release the carbon dioxide can be obtained by multiplying the ratio, 0.044, by the annual release of carbon in carbon dioxide, 5.4 gigatons. This indicates an emission of methane of 0.22 gigatons of carbon in methane. The value of 0.22 gigatons is 8 times higher than the estimate [2] of 0.03 gigatons of carbon in methane.

How credible is the new estimate of methane release to the air from processes associated with combustion? The critique will be offered on two levels: general critical arguments and other kinds of information.

First, the correlations in Figures 1 and 2 can, at best, only argue for sources of the two gases originating in the same areas, not that both must derive from the same combustion source. But, of course, combustion is the source of the carbon dioxide. Thus,

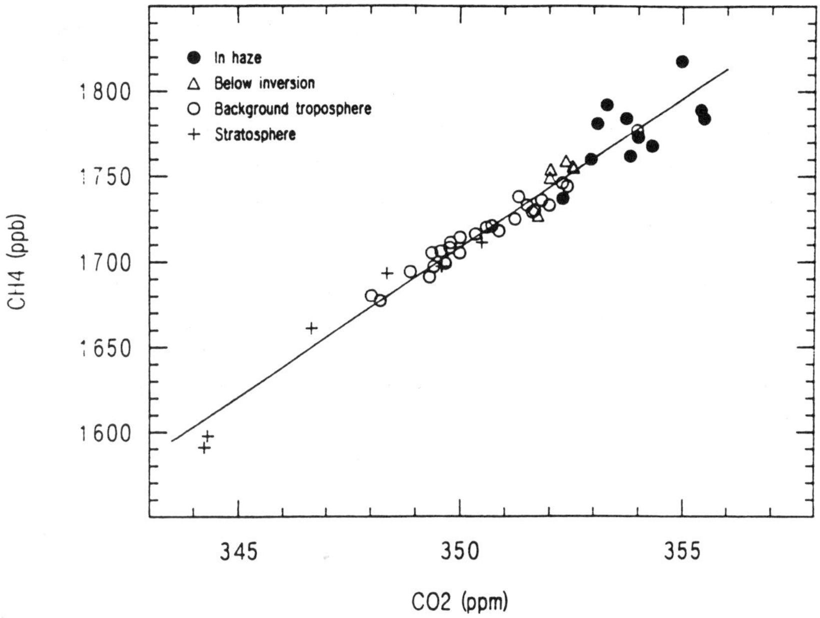

FIG. 1. The relationship between simultaneous samples of carbon dioxide and methane taken aboard aircraft in April 1986 in the Arctic. The slope of the best fit line is 17.5 ppb of methane (CH_4) per 1 ppm carbon dioxide (CO_2) and the correlation coefficient is 0.97. (Reproduced from Conway and Steele [6])

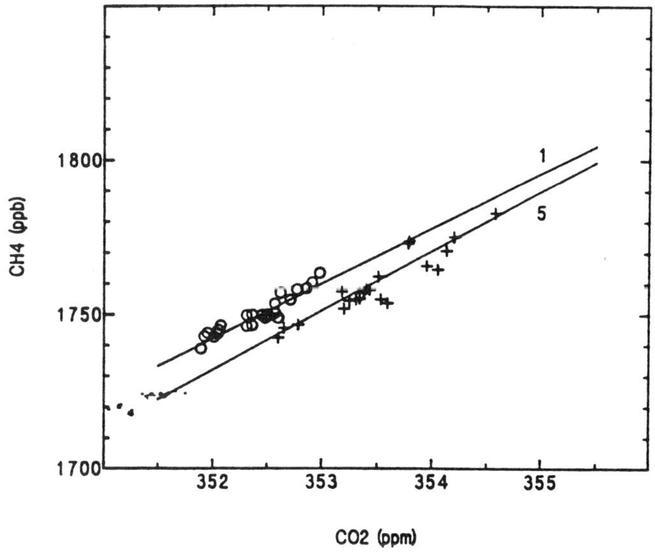

FIG. 2. The relationship between simultaneous hourly samples of carbon dioxide and methane in ground level air at Point Barrow, AK, during March-April 1986. The circles and crosses each refer to separate periods. The slopes of the best fit lines are 0.018 and 0.019 ppb of CH_4 per 1 ppm CO_2 and the correlation coefficients are 0.93 and 0.94. (Reproduced from Hansen et al. [7])

short of monitoring stacks or the nearby areas around the combustion sources, the assumption that methane comes from the same stack or tailpipe as the carbon dioxide, e.g., combustion, remains unproven. Further, even if the Arctic area sources of carbon dioxide also release the methane, one is not justified in assuming that they would do so everywhere in the world. On both of these scores, this new, much higher estimate of direct anthropogenic methane release remains questionable and, if anything, overestimates the methane release from combustion.

There is other information that supports the higher man-made methane releases. NOAA has also measured, simultaneously, the methane and carbon dioxide concentrations at Boulder, Colorado [6]. Here, as well as at Point Barrow, Alaska, there exists a positive relationship between the two gases but with lower correlation coefficients and a smaller ratio of methane to carbon dioxide change (0.0076 compared to the Arctic value of 0.018 used above). This is to be expected since the air sampled at Boulder has passed over biogenic sources and sinks of both gases since last passing over man-made combustion sources. A combustion-methane signal, if any exists, would be expected to be weaker than at Point Barrow in the spring. It has also been shown [7,9] that a positive correlation occurs between samples of carbon black and carbon dioxide, both of which are known to have combustion sources. In this case, the ratio of the two sampled substances fits reasonably in the range of various likely combustion sources, another example of the benefit of simultaneous sampling of air qualities.

Second, and more significant, there are recent measurements of ^{14}C in methane. Lowe et al. [10], have measured the radiocarbon in clean Southern Hemisphere air from which they estimate that 32% of the carbon in methane is "dead" and therefore of fossil origin. The estimate of direct anthropogenic methane is 7.5% of their estimated total methane source [2]. Thus the ^{14}C data suggest a factor of 4 more methane in the air from fossil fuel combustion than Khalil and Rasmussen. The factor of 4 gives the same sense of increase as the value 8 derived from the simultaneous measurements and assumptions. They both are indicative of much more man-made methane released to the air than otherwise estimated.

The sense of this report is the benefit that can be achieved by simultaneously measuring many substances at the same time and place. It is illustrated by an interpretation leading to more combustion-associated methane emissions than reported in the earlier literature.

References

1. P. Steele, et al., *J. Atmos. Chem.* **5**, 125-171 (1987).

2. M.A.K. Khalil, and R.A. Rasmussen, *J. Geophys. Res.* **88**, 5131-5144 (1983).

3. D.R. Blake, and F.S. Rowland, *J. Atmos. Chem.* **4**, 43-62 (1986).

4. P.J. Fraser, et al., *J. Atmos. Chem.* **4**, 3-44 (1986).

5. H. Craig, and C.C. Chou, *Geophys. Res. Lett.* **9**, 1221-1224 (1982).

6. T.J. Conway, and L.P. Steele, *J. Atmos. Chem.*, in press (1988).

7. A.D.A. Hansen, et al., *J. Atmos. Chem.*, in press (1988).

8. R. Rotty, personal communication (1988).

9. J.D. Kahl, and A.D.A. Hansen, *J. Atmos. Chem.*, in press (1988).

10. D.C. Lowe, et al., *Nature* **332**, 522-525 (1988).

NITROGEN OXIDES AND OZONE

Ralph M. Perhac

Electric Power Research Institute
P.O. Box 10412
Palo Alto, CA 94303

The major air quality environmental concern over the past decade has been acid rain, both sulfur dioxide (SO_2) and nitrogen oxides (NO_x) emissions playing an important role. Most attention, by far, has focused on SO_2. The importance of nitrogen oxides in the environment, however, should not be overlooked because they bear not only on the acidic deposition issue but play an important role in another environmental concern—tropospheric ozone. Many metropolitan areas are in a state of nonattainment, hence the ozone issue is one of considerable concern to society and, particularly, to the electric utility industry. As recently as 1987, EPA designated 65 urban areas as nonattainment for ozone, and those 65 have an estimated population of nearly 130 million people.

Ozone is a health concern, and humans appear to be particularly sensitive to ozone exposure during exercise. The U.S. Office of Technology Assessment estimates that about 80,000 people are exposed to levels of ozone exceeding 0.12 ppm (the U.S. ambient standard) while doing heavy exercise, for an average of about four hours per person per year. The population exposed to concentrations exceeding the standard increases to 13 million when performing moderate exercise, for an average of about six hours per person per year, and 21 million are estimated to be exposed to levels exceeding the standard while performing low exercise, for an average of eight hours per person per year. Ozone reduction in nonattainment areas is a matter of high concern.

The utility industry is concerned over health consequences. But it is also concerned over proposals for more stringent NO_x legislation which would be costly to the industry and to consumers but which might do little to alleviate the NO_x/ozone issue and, in fact, might exacerbate it. The chemistry of ozone formation is such that a simple reduction of the NO_x precursor does not necessarily insure a reduction in atmospheric ozone.

Figure 1 is an overly simplified picture of the atmospheric chemistry of ozone formation. Despite the simplification, it does point out the importance of volatile organic compounds (VOC) in allowing accumulation of ozone.

Without VOC, a steady NO_2–NO–O_3 state is reached, depending on sunlight intensity. Obviously, the ratio of VOC to NO is critical in defining the ultimate atmospheric

Copyright 1989 by Elsevier Science Publishing Company, Inc.
Global Climate Change Linkages
James C. White, Editor

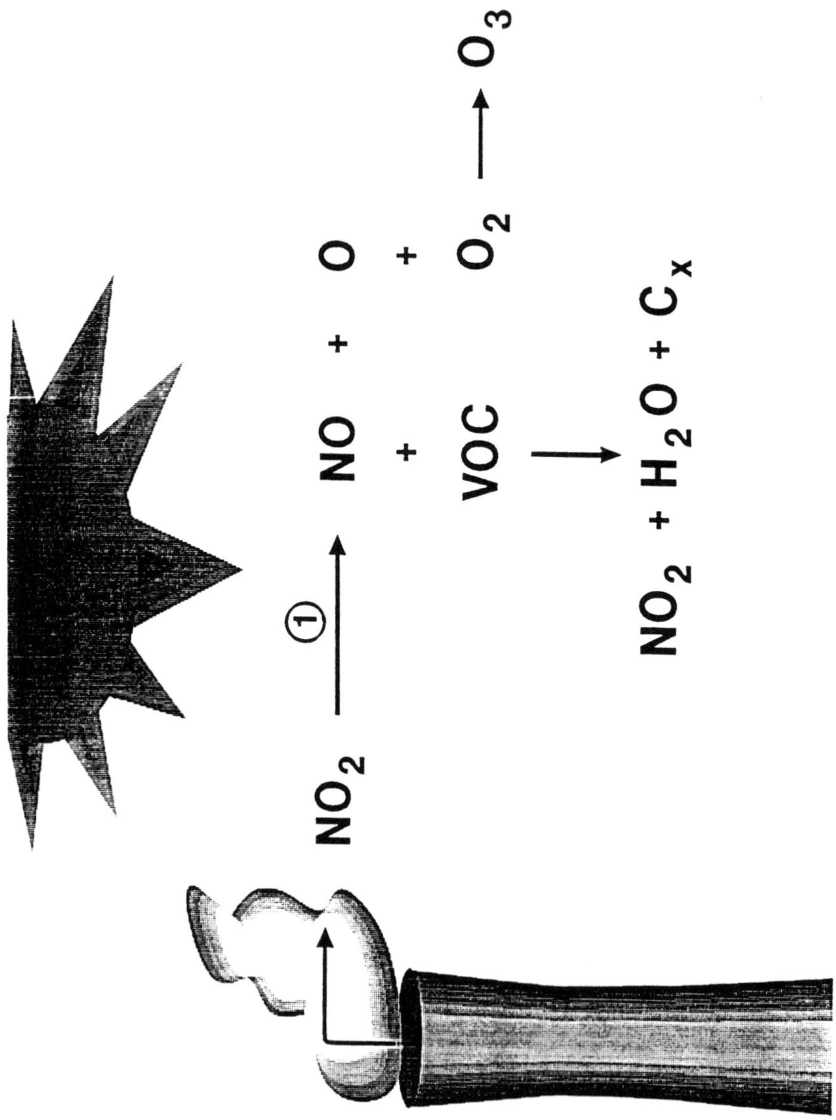

FIGURE 1

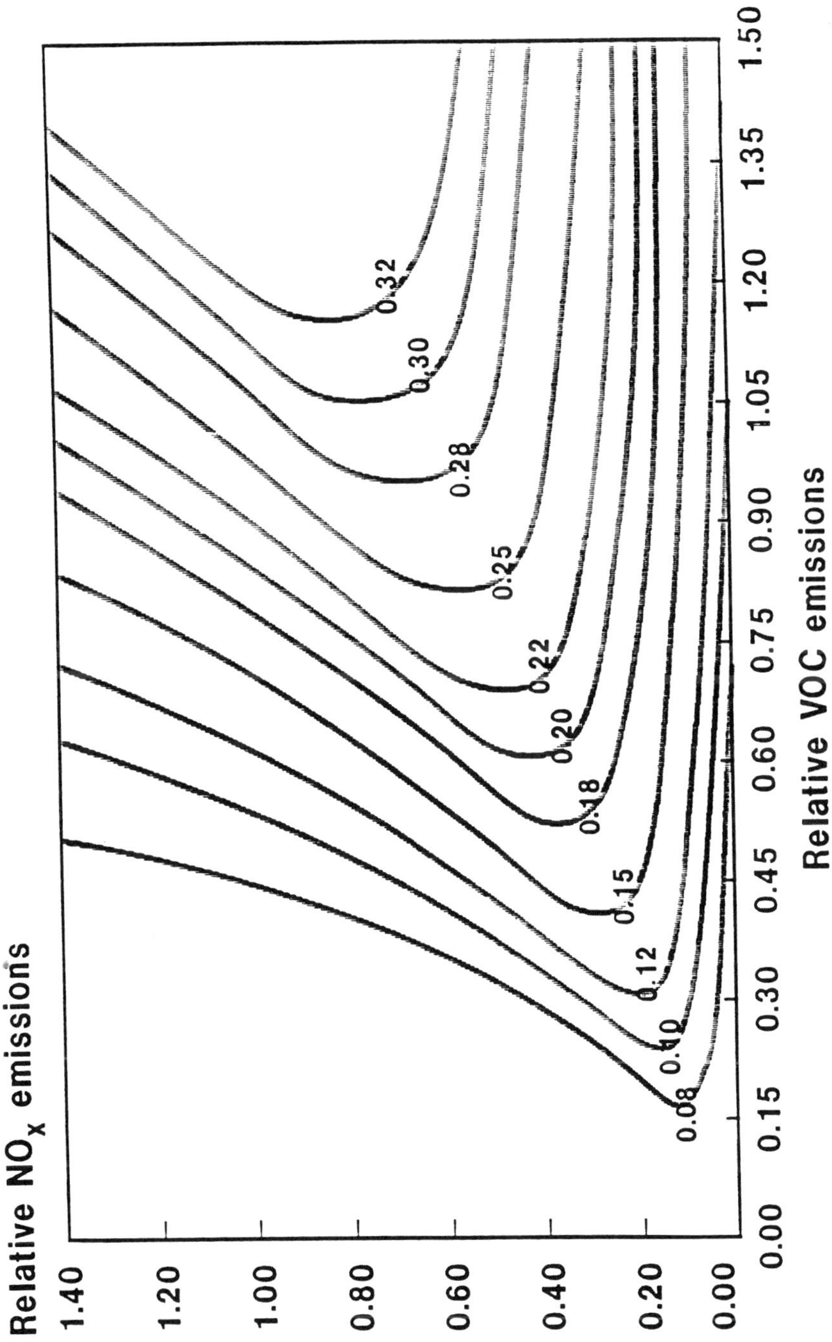
Relative NO_x emissions
1.40
1.20
1.00
0.80
0.60
0.40
0.20
0.00
0.00
0.15
0.30
0.45
0.60
0.75
0.90
1.05
1.20
1.35
1.50
Relative VOC emissions
0.08
0.10
0.12
0.15
0.18
0.20
0.22
0.25
0.28
0.30
0.32

FIGURE 2

concentration of ozone. High VOC:NO ratios will favor ozone formation. Conversely, high NO:VOC ratios favor attainment of steady state. But in the steady state situation, reduction of NO_x emissions could increase the role of whatever VOC is present, thereby actually leading to an increase in ozone. The situation is not simple. Reduction in VOC will almost always result in ozone reduction; reduction in NO_x may or may not (Figure 2).

Recently, EPRI decided to evaluate the efficacy of NO_x control in reducing ozone in an urban area. To do so, it applied the EPA Urban Airshed 4 model to Philadelphia, Pennsylvania, a nonattainment urban area. Data on air quality and meteorology came from a July 1979 study conducted by EPA. Detailed emission inventories for hydrocarbons and NO_x were included in the exercise. The model was run against observed (1-hour average) ozone concentrations at ground level for a test-case day (July 19, 1979) – the base case. The model was run for four emission reduction "scenarios" with results being compared to the base case. The four "scenarios" were:

S1 – 25% reduction of only VOC emissions
S2 – 25% reduction of VOC plus 25% reduction of power plant NO_x
S3 – 25% reduction of VOC plus 25% reduction of all NO_x
(mobile and stationary sources)
S4 – 25% reduction of only power plant NO_x emissions

The modeling results (in summary) are as follows:

S1 – ozone decreased by 0.01 – 0.02 ppm
S2 – ozone increased by 0.01 – 0.02 ppm
S3 – ozone increased in some areas, decreased in others
S4 – ozone increased by 0.01 – 0.03 ppm

Obviously, reduction in NO_x yields uncertain improvement in ambient ozone concentration. VOC reduction alone leads to decreased ozone. NO_x reduction alone leads to increased ozone. Reducing both NO_x and VOC leads to more complicated results.

The above-cited example of a model exercise is not cited as a reason for avoiding NO_x control. It is given only to caution that any NO_x control should consider the complexities of atmospheric chemistry as an important part in judging what results will be achieved.

Policy Linkages*

John Hoffman

U.S. Environmental Protection Agency
Office of Air and Radiation (ARN-445)
401 M Street SW
Washington, DC 20460

Current technology is considered to be inadequate and too expensive to solve problems related to climate change. Unless we improve our approach, we will be unhappy with the outcome.

We must ask if the technology that is developing is appropriate; do good solutions exist? Are we making sufficient investment in technological change? Are we removing obstacles and inhibitions that society places on change? We must cease to look on atmospheric problems in terms of dollars alone. Are we mobilizing talent to solve the problems?

The outlook on tropospheric ozone is dismal. We have regulations which tell a company what to do, and they comply, but there is little incentive to develop technology to further reduce the amount of emissions. There are no profits to be made by reducing them. Companies may be deterred from innovation by the thought that new technologies bring new regulations, and agencies are looking for new areas to control.

There are many problems with stratospheric ozone and they are likely to get worse. We are not spending much money in this area, compared to other environmental problems. Few people are working in this area and progress is slow.

With stratospheric ozone, the early agreements to limit aerosol propellants gave us some time, but there have been few technological changes since. When it became apparent that further regulations were not forthcoming in the 1980s, companies stopped searching for substitutes. The Montreal Protocol limits production of CFCs, which will drive up the price and encourage innovation.

Hidden regulations are a hindrance to progress, effectively slowing the search for acceptable substitutes. For example, military specifications for electronic cleaning solution specify CFC 113 only; nobody has incentive to find a substitute. We are working with the military to correct that. We expect 20 substitutes within a year, probably at lower

*This paper was taken from a tape recording of Dr. Hoffman's remarks made at the conference. Any errors it contains are the responsibility of the editor.

Copyright 1989 by Elsevier Science Publishing Company, Inc.
Global Climate Change Linkages
James C. White, Editor

cost. Underwriters Laboratories require that refrigerants be nonflammable; this restricts research on compounds other than Freons, compounds and mixtures which can possibly be rendered safe and also hold promise to be much more efficient with proper usage. There is difficulty finding money in some firms for development. They would rather accept present products, get an exemption or do nothing.

Firms are reluctant to invest in the research and development necessary to develop substitutes. People are not necessarily using the best alternatives. They use any unregulated chemicals available, regardless of energy efficiency.

To meet the greenhouse problem, we need long-term research. Most firms want payback in about three years for any innovations made, which does not encourage long-term research. Firms are not concentrating on possible markets for energy conserving products and processes. Very little is being done on non-fossil fuel based energy production.

We need long-term progress, ready financing, raised motivation for investors and companies, multiple payoffs and more talented researchers. Few people in this country are concentrating on problems like refrigeration. The innovations are coming from Germany and Japan.

If we don't act now, we'll pay later. We will find ourselves with high energy use and low efficiency, competing with countries which have overcome these problems.

We will not make much progress without adding resources and getting our most talented people to work on the problem.

ENVIRONMENTAL PROGRAMS DURING AN ERA OF DEFICIT BUDGETS

Lester B. Lave

Graduate School of Industrial Administration
Carnegie-Mellon University
Pittsburgh, PA 15213

Environmental pollution presents difficult scientific problems because it is caused by the discharge of "residuals." No one wants to cause these environmental problems; rather they are the result of disposing of materials after they have been "used up" (Kneese, Ayres and d'Arge, 1970).

The residual nature of the emissions means that no one thought of urban ozone problems when they first designed automobile engines and fuels; no one thought about acid rain when they were attempting to bring the benefits of cheap electricity to all Americans; certainly, no one thought about global climate change when deciding to clear forest in order to grow crops. Even for global problems, the offending discharges are tiny (relative to the scope of the activity) and are often forgotten. Often the extent of the resulting problem cannot be estimated in advance.

Our inability to foresee all the consequences of our actions is a reason to take care before acting. This is not a plea for inaction, but rather a plea for careful thinking before "corrective" actions are undertaken. Without careful evaluation of the proposed actions, they may not solve the problems; indeed, the actions might cause worse problems.

Great Britain has had acid rain problems for more than a century, according to Park (1988), a geographer at the University of Lancaster. In 1852, Robert Angus Smith linked the sooty skies over Manchester with acidity of precipitation; he coined the term "acid rain" in an 1872 book.

Environmental Perspectives

The most important force shaping action is the individual's environmental perspective. Some people don't think the USA has any important environmental problems; some see problems but are willing to spend little or nothing to solve them; some see important problems but very limited resources for solving them; and some believe the only scarce resource is public attention.

Copyright 1989 by Elsevier Science Publishing Company, Inc.
Global Climate Change Linkages
James C. White, Editor

The Frontier or OMB Perspective

To a frontiersman, a forest had to be cut and a prairie plowed in order to make the land suitable for habitation (Hays, 1987). The natural state of the environment was a challenge to be overcome.

The Office of Management and Budget (OMB) is the institutional curmudgeon of government. OMB is the force for not spending resources to solve problems, or for spending as little as possible. In 1980, David Stockman gave one of the purest expressions of this view: "How much are the fish worth in those 170 lakes that account for 4 percent of the lake area of New York? And does it make sense to spend billions of dollars controlling emissions from sources in Ohio and elsewhere if you're talking about a very marginal volume of dollar value either in recreational terms or in commercial terms?" (as quoted in Regens and Rycroft, 1988). In other words, what is the dollar value of a trout in an Adirondack lake, a trout that will never be caught or eaten by any person?

An Economic or Conservationist View

Around the turn of the century, the conservationist movement rejected the frontier viewpoint that the only good tree was a felled tree. They saw the environment as providing services to people: lumber, minerals, recreation, pasture, and water. They wanted to slow or stop forest clearing, stop erosion, and set aside national parks that could be used for providing timber, pasture land, minerals, and recreation. Their desire was not to preserve acres of virgin timber, but instead to use the land fully for the benefit of the economy and people.

Economists come into the picture by designing programs that "maximize" the productivity of the environment. Economists want us to stop the squandering of our environmental resources and start developing them for the greater benefit of society.

From an economic perspective, the amount of money the federal government has available for environmental programs is limited. Thus, the emphasis should be on allocating the funds among programs so that the largest gain in environmental quality is achieved. The techniques for doing so are classified as "cost-effectiveness" or "benefit-cost" (the terms are not equivalent).

Preservationists

The third perspective refers to those who want to preserve the environment in its natural state. Use by people is not the goal, and indeed people shouldn't be permitted to use certain areas if their use would destroy or even change the environment. Thus, fragile environments ought to be off-limits to hikers and much wilderness ought to be used by hikers only under highly regulated conditions. They reject the conservationists' ideas of achieving the fullest use of the environment and instead want to preserve what is there.

Preservationists believe that environmental resources are limited (many sites are unique and we need to stop species extinction) but that resources for the environment aren't. The largest problem is to alert the public to the damage being done to the environment; a concerned public will stop the despoliation.

I have a simple test of which category you fit into: What do you call a body of shallow water that is a breeding place for mosquitoes? If the name "swamp" springs to mind, you have a frontier perspective. If the name "wetlands" was your answer, you are in one of the other categories. If you praise God for making mosquitoes because they are food for young fish, you are a true preservationist.

Environmental History

The frontier mentality tended to dominate the USA until the late 1960s. The stronger version of this perspective tended to disappear toward the end of the last century, when clearing a forest ceased to be regarded as a good thing, independent of what was being done with the timber or land. News reports indicate that Brazil has only recently stopped giving incentives to clear tropical forest. Conservationist voices began to be heard around the turn of the century, but that view was not dominant. There are still people with the OMB viewpoint, but it has ceased to dominate public or political thinking.

The environmental movement bloomed in the 1960s: Rachael Carson's *Silent Spring* was a landmark. By Earth Day in 1969, there had been major sensitization of the public to environmental problems and the need to end uncontrolled dumping. The view was largely conservationist, with people being angry at air pollution in cities, water pollution in rivers, and the dumping of toxic waste. By 1970 or 1972, some Congressional attitudes (as expressed in legislation) were preservationist. The Endangered Species Act, requirements for environmental impact statements, and declaration in the Clean Water Act of 1972 that the goal was to end discharges in waterways by 1985 all portray this preservationist perspective. The legislation and its implementation were opposed by some polluters, who did not feel that the government should or could control their use of land and discharges. However, both in the political and legal arenas, these objections were swept aside.

The Vietnam War had many costs; one that is usually not recognized is that environmental programs were a victim of the budget reality forced on the nation at the end of the war. The rhetoric of preservationism prevailed through the 1970s, but the reality was that the budgets of EPA and other regulatory agencies did not grow sufficiently to implement the new responsibilities.

The early Reagan administration saw a brief return to the frontier perspective, but this reversion was so far out of line with Congressional and public opinion that it disappeared with Ann Gorsuch-Burford (Lash et al, 1984). The remainder of the administration reflected a conservationist perspective, with very limited budgets for environmental programs. The past seven years have been times when public pressure managed to restore some of the budget cuts from the early 1980s, but EPA was faced with deciding how to allocate a limited budget among a large number of programs.

In my judgment, the Bush administration is going to continue this trend. On the one hand, President Bush is more attuned to public concerns. On the other hand, he faces more stringent pressure to cut the deficit and claims he will not raise taxes and will not cut Social Security and will keep defense expenditures high. If so, large cuts in domestic programs are required and it is hard to see that environmental programs will be exempt. Thus, the Bush administration will require the environmental community to fight hard to maintain viable programs. Increasing expenditures on these programs seems an impossible dream.

If so, the theme for environmental programs is going to have to be to focus on each program and figure out how to use the resources as productively as possible (Raufer and Feldman, 1988). Hard decisions are going to have to be made about, for example, taking resources from cleaning up toxic waste dumps in order to have resources for dealing with greenhouse effects. This prediction is certainly not meant to be pessimistic or to discourage environmentalists from increasing pressure on Congress and the new administration. However, intelligent planning requires attention to budget realities.

Scientific Uncertainty About Causes and Effects

Large amounts of uncertainty remain about the processes leading to urban air pollution and its effects; there is still greater uncertainty about acid rain, and then much greater uncertainty about greenhouse issues. Statements made a decade ago that forests and lakes have suffered major damage from acid rain have had to be retracted, as have the statements that sulfur dioxide emissions are the cause of the problems. It is not that these statements have been shown to be untrue, simply that each was a preliminary conclusion that was contradicted by new data. "Complexity is the norm for natural ecosystems..." according to Gene Likens (1988), the chronographer of Hubbard Brook.

Similarly, the effects of acid rain on trees have had to be reexamined. According to Kulp (1988) of EPA, experiments show that ozone decreases the yield of crops and produces effects in trees like those ascribed to acid rain. In contrast, "SO_2 and NO_x may be dismissed from serious consideration as having major impacts on vegetation except in some local situations..." Furthermore, acid precipitation between pH 4.0 and 5.0 has no measurable effect. Acidity must be increased below pH 3.0 before effects begin to be observed. Indeed, deposition of NO_x helps plants. The picture is less benign for high elevation forests, since they are stressed by the environment and thus more susceptible to other stressors, such as ozone and acid deposition. Kulp conjectures there is a multiplicity of agents responsible for forest dieback – from ozone to direct exposure to acid rain to acids leaching nutrients from soil and mobilizing heavy metals and especially exposure to hydrogen peroxide. The solution is not as simple as curtailing sulfur emissions.

The obvious answer is to embark on a large research program to get answers to the critical questions. However, research not only improves scientific understanding, it is a way of delaying action. When the acid rain controversy heated up, a chorus of Senators and Congressmen from coal burning states began to sound like the president of the National Academy of Sciences: only the best science was acceptable and research programs should be launched to get the answers. However, not all had been converted to "good science;" Regens and Rycroft point out that the Reagan administration called for more research – while cutting the EPA research budget.

Conclusion

Uncertainty, and even misinformation, are important components in determining policy. For example, Park (1988) recites the conventional criticism of Great Britain for being the worst polluter in Western Europe, since it has the most sulfur emissions. However, a fairer measure of contribution to acid rain is the level of emissions per capita, since there is no reason to blame a large country and exonerate a small one just because of size. Using this measure, Britain has less emissions per capita than Denmark and Sweden and is tied by Belgium. Thus Britain's guilt must be interpreted in light of the greater guilt of two of the Scandinavian countries.

Environmental groups made the judgment in the early 1980s that public concern for acid rain was so great that they could bundle acid rain control with the Clean Air Act revisions and force the latter. That judgment was wrong, since it resulted in neither a revised Clean Air Act nor effective legislation on acid rain. Given the high level of public concern about greenhouse effects, it is tempting to try to use this concern as the lever to achieve effective legislation on air pollution, acid rain, and greenhouse effects. In my judgment, this bundling is misguided. The price tags are formidable for dealing with tropospheric ozone problems, with small particles, toxic air pollutants, reducing emissions of sulfur and nitrogen oxides, and curtailing the emissions of greenhouse gases. Combining all these concerns into one mammoth bill presents a formidable price tag.

The most attractive legislation would be that which abates a pollutant that helps manage two or more of these issues simultaneously. For example, urban ozone has emerged as the most important urban air pollution problem. At the same time, there is growing consensus that it is ozone, not acid deposition, that causes the damage done to trees by acid rain. To a small extent, this tropospheric ozone is also a greenhouse gas. Thus, ozone control emerges as a central issue in two problems and as a concern in the third. Ozone control should provide a major focus for environmental regulations.

Some of the policies aimed at curing one problem have only exacerbated another. For example, tall stacks were viewed as a quick, inexpensive solution to lowering the concentration of sulfur dioxide around the plant. However, the tall stacks increased the magnitude of the acid rain problems. A great deal of attention needs to go toward resolving the scientific uncertainty to learn the best ways of managing environmental problems and to ensure that current solutions don't lead to problems elsewhere.

Thus, both because of scientific uncertainty and budget limitations, these environmental problems should be attacked thoughtfully. As uncertainty about effects is resolved, uncertainty about how to lower emissions must be resolved; the public, government agencies, and Congress should be in a position to react swiftly to the new information. For now, there are many programs, from energy conservation to emphasizing natural gas rather than coal, to controlling ozone, that are compelling and should have high priority.

References

Hays, Samuel P., *Beauty, Health, and Permanence: Environmental Politics in the United States, 1955-1985*, Cambridge: Cambridge University Press, 1987.

Kneese, Allen V., Robert U. Ayres, and Ralph C. d'Arge, *Economics and the Environment: A Materials Balance Approach*, Baltimore: Johns Hopkins University Press, 1970.

Kulp, J. Laurence, "Source-Receptor Relationships: Information Needs--Terrestrial," 95-100, in White (1988).

Lash, Jonathan, Katherine Gillman, and David Sheridan, *A Season of Spoils: The Reagan Administration's Attack on the Environment*, New York: Pantheon Books, 1984.

Lave, Lester B., "The Greenhouse Effect: What Government Actions Are Needed?" *Journal of Policy Analysis and Management*, **7**, 460-70, 1988.

Likens, Gene E., "Source-Receptor Relationships: Information Needs – Aquatic," 101-120, in White (1988).

Park, Chris C., *Acid Rain: Rhetoric and Reality*, London: Methuen, 1988.

Raufer, Roger K. and Stephen L. Feldman, *Acid Rain and Emissions Trading: Implementing A Market Approach to Pollution Control*, Totowa, N.J.: Rowman and Littlefield, 1988.

Regens, James L. and Robert W. Rycroft, *The Acid Rain Controversy*, Pittsburgh: University of Pittsburgh, 1988.

White, James C. (ed.), *Acid Rain: The Relationship between Sources and Receptors*, New York: Elsevier. 1988.

THE ENVIRONMENTAL ATTRIBUTES OF NATURAL GAS

Michael I. German

American Gas Association
1515 Wilson Boulevard
Arlington, VA 22209

Abstract

In many nations, increased reliance upon natural gas resources in national energy planning offers opportunities to simultaneously minimize pollution, including the build-up of "greenhouse gases," while also choosing a high efficiency energy strategy. Natural gas resources are abundant, widely distributed, and underutilized. By all measures— air emissions, solid wastes, water pollution, and water consumption—natural gas is by far the most environmentally desirable fossil fuel. Compared to other energy alternatives, natural gas systems are also highly capital and resource efficient, requiring significantly lower capital investment to produce, deliver and utilize a unit of energy. Recently commercialized utilization technologies have greatly enhanced these advantages.

In electricity generation, natural gas is emerging as the fuel of choice. Natural gas-fueled electric generation offers lower capital and operating costs, much shorter construction lead times, high efficiency, rapid start-up, increased reliability and minimum environmental impact. Gas-fired electric-generating equipment also provides planning flexibility by easily accommodating incremental capacity additions that can track electric load growth more closely. In many applications, natural gas can complement coal use by increasing its efficiency and significantly reducing emissions.

Introduction

The environmental attributes of natural gas have long been recognized. The use of natural gas produces far lower emissions than the use of other fossil fuels, including emissions of the most commonly identified "greenhouse gases"—carbon dioxide, nitrogen oxides, methane and CFCs. Pollution control can be quickly achieved using natural gas, since many emitters already have gas service. The use of natural gas entails significantly lower capital investment and operating costs than other fossil fuels, and only natural

Copyright 1988 by the American Gas Association

Published by Elsevier Science Publishing Company, Inc.
Global Climate Change Linkages
James C. White, Editor

gas offers the ability to reduce emissions of the entire spectrum of pollutants simultaneously.

These environmental advantages have been taken advantage of only to a very limited degree. Why? First, until recently, concern about environmental pollution has been regarded as a special problem – almost a luxury – of the most developed nations. Second, even where air pollution has been a consideration – as in the United States – preconceptions regarding the future price, supply and technology for natural gas resulted in gas use being precluded or ignored as a compliance method. Today these views are changing.

Air quality problems are obvious in most of the world's cities. If anything, these problems are worse in developing nations, many of which have a heavy reliance upon diesel-fueled vehicles. Solid wastes, water pollution and water consumption problems are becoming recognized as equally widespread and serious, if not always as visible.

The truly transnational character of environmental degradation has also become recognized. The impacts of many forms of pollution know no international boundaries. For example, in the past decade considerable attention has been given to such topics as ocean pollution and "acid rain." Climate change, while not a new issue, is the example which has most recently achieved a high public profile.

Questions of science aside, the transnational pollution issues are, in many respects, the most politically contentious, since the impacts of pollution may not be felt proportionally in the nation of origin. It is a challenge to the international community to find the mechanisms to control pollution, when the nation that must pay for environmental control may not be the direct and immediate beneficiary of such control or where the impacts may not be obvious for some years.

Another key factor contributing to the difficulty of pollution control is the concept that pollution control is costly. Clearly, the high capital cost of many forms of control is an impediment, especially for developing nations which are already under serious burdens of international debt. Current developments in the natural gas industry and in gas technology offer an almost serendipitous solution. In many nations, increased reliance upon natural gas resources in national energy planning offers opportunities to simultaneously minimize pollution, including the build-up of "greenhouse" gases, while also choosing a least-cost energy strategy. That is, reducing pollution through increased use of natural gas – the cleanest fuel – may actually reduce the cost of providing and using energy.

Much of energy policy making worldwide has been driven in recent years by the incorrect belief in a few basic tenets: among them, that oil and natural gas would be in increasingly short supply, especially in the United States; that oil and natural gas prices would inflate rapidly; that, in spite of these developments, world economic growth would be steady and robust; that central station coal and nuclear fission electric plants would be a highly cost-effective source of energy in this century; and that renewable electric energy would become cost-effective before the turn of the century, with fusion energy to follow. A corollary tenet has been that natural gas is a premium fuel and feedstock, whose highest benefit to producer nations could be achieved through export, either as liquefied natural gas (LNG) or through conversion into such products as ammonia fertilizers and methanol. [1]. All of these basic tenets have now been shown to be, in varying degrees, incorrect.

Globally, oil and natural gas are in excess supply both worldwide and in the United States. Natural gas supplies are likely to be adequate for the rest of this century and most of the next century. A variety of new natural gas supply and utilization technologies has emerged. These developments should guarantee the availability and competitiveness of natural gas. The rapidly escalating capital costs of nuclear and coal-fired units, as well as associated environmental and regulatory barriers, have diminished the attractiveness of these options. In addition, the demand for energy has become less predictable due to conservation, increasing uncertainty in the business cycle, and other factors. Finally, the high capital costs and market fluctuations associated with the LNG, ammonia, and methanol trades have meant that netbacks to producer nations from these downstream investments often have been less than satisfactory.

Large-scale energy projects with high capital costs and long lead times have shown themselves to be increasingly unworkable in a late twentieth century environment of uncertainty and rapid change, particularly when the projects involve international trade. Future planning will place a premium on energy alternatives that offer short lead times, low capital costs per unit of energy, and the ability to add capacity in small step-wise increments. Environmental and public health impacts will also be an increasingly important consideration as population and energy consumption grow.

These developments necessitate a reconsideration of the role of natural gas in energy policy and development. While international trade will certainly play an important role, greatly increased domestic utilization of gas resources will be desirable in many countries.

Perhaps the most profound implications are for electric generation. For this application, natural gas may be emerging as the fuel of the future. Natural gas-fueled electric generation offers lower capital and operating costs, much shorter construction lead times, high efficiency, rapid start-up, increased reliability and minimal environmental impact. Gas-fired electric-generating equipment also provides planning flexibility by easily accommodating incremental capacity additions that can track electric load growth more closely. In many applications, natural gas can complement coal use by increasing its efficiency and significantly reducing emissions.

Natural gas offers value directly to electric utilities in the form of greater efficiency, environmental enhancement and operating superiority. Indirectly, natural gas can provide low-cost electricity to utilities through cogeneration and independent power production.

World Natural Gas Resources and Consumption

There are three levels of gas and oil resources: proved reserves; economically recoverable resources; and in-place or total resources (see Exhibit 1). The first category already has been discovered and is considered producible under current economic and operating conditions; the second category is believed to exist and estimated to be producible under economic and operating conditions which are presently considered likely to exist in the foreseeable future; the third category is the total resource believed to exist in all deposits, of which some portion (depending on technology and economics) could be produced. With time and technological development more and more of this gas will become economically recoverable. [2].

As a part of the in-place resources, there are sources, other than conventional, of both gas and oil that require further research, improved technology and/or higher energy prices before they can be developed in significant amounts. The term "nonconventional," for the purposes of this paper, refers to oil shale and tar sands for oil, and tight sands, Devonian shale, coal seam methane and enhanced gas recovery for gas. The existence of these sources is well documented and estimates are based on numerous known occurrences worldwide.

However, there are two potential sources of gas which differ from the other nonconventional sources because of their intangible nature and the vast quantities of energy which could result (if theories supporting their existence are proved to be correct). These are methane from gas-hydrates and abiogenetic methane or deep-earth gas. As more knowledge of these sources is gained, there is the potential for extremely large quantities of gas being made available worldwide, even at very low levels of recoverability. The theories surrounding deep-earth gas are at present unproved and, while the existence of gas-hydrates is acknowledged, very little is known about this potential gas source.

When all three levels of resources are considered and compared to current consumption, there are clearly sufficient gas resources in place worldwide to satisfy any reasonable demand requirements through the twenty-first century and beyond (see Exhibit 2). Further, it is widely recognized that many known gas reservoirs are currently untapped, often in nations which rely heavily upon imported petroleum or refined petroleum products.

At the 16th World Gas Conference in Munich, Germany, in 1985, the International Gas Union (IGU) released its Report of the Task Force on World Gas Supply and Demand, 1983-2020. In this report the IGU projected potential world natural gas production and consumption on the basis of foreseeable economic conditions. In other words, the IGU task force investigated how much of the gas resource base might actually be brought to market in the 1983-2020 time period. [3].

Total potential supply of pipeline quality gas projected by the IGU study is shown in Exhibits 3 and 4. As shown, in 2020 potential world gas production is projected to be

Exhibit 1
World Resources Triangles
(not exactly to scale)

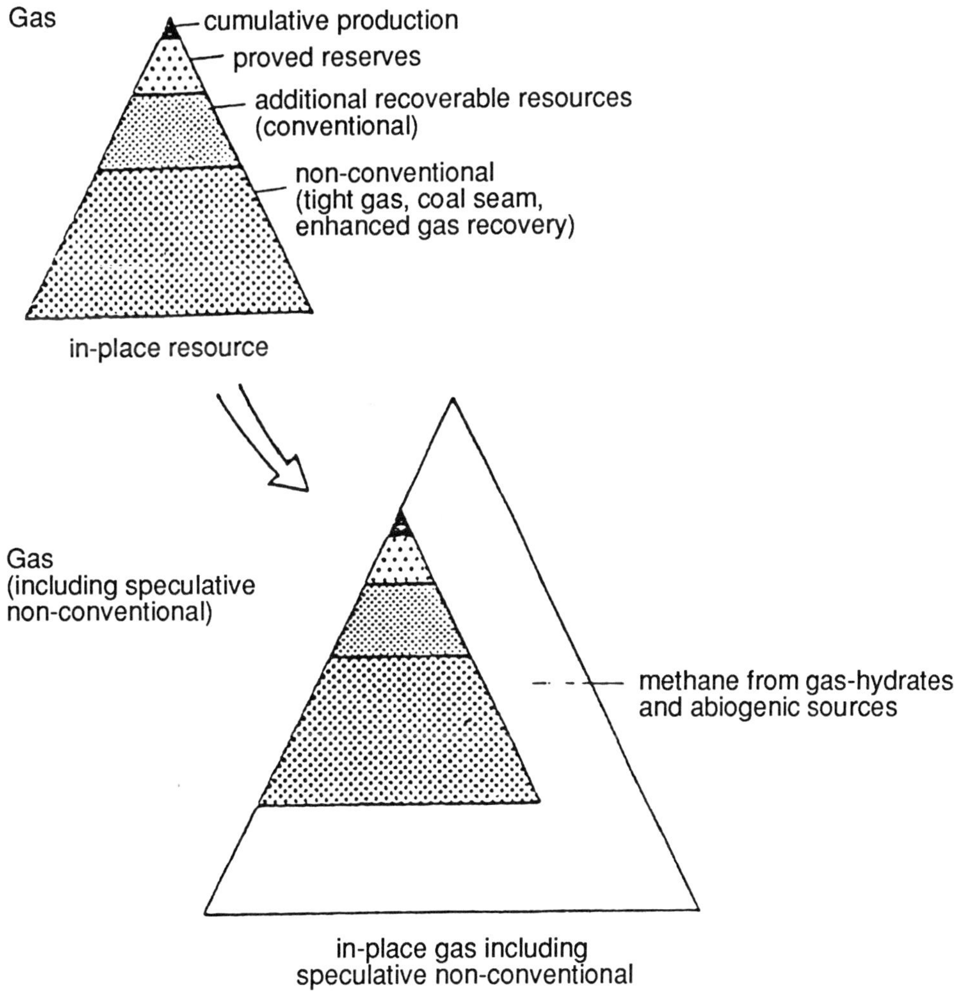

Exhibit 2

World Gas Consumption Versus Resources2/
(Quadrillion Btu-1015)

1983 Consumption	Proved Reserves	Remaining Recoverable	In-Place Resource (incl. non-conven.)	In-Place Resource (incl. speculative non-conven.)
55	3,270	9,430	143,000	271,143,000

Exhibit 3

International Gas Union

Total Potential Gas Supply
(quads)*

	1983	2000 M	2000 R	2020 M	2020 R
Conventional natural gas	55	91	101	103	114
Unconventional gas	_-	_5	__9	_10	_15
Total	55	[97]	110	113	129

M = Moderate Oil Price Rise. Oil price reaches $55/bbl in 2020 (1983 dollars).
R = Rapid Oil Price Rise. Oil price reaches $70/bbl in 2020 (1983 dollars).
* Original figures were reported in exajoules. One exajoule equals approximately one quadrillion Btu or one "quad".
[] SIC

Exhibit 4

International Gas Union

Potential Supply of Commercial, Conventional & Unconventional Gas

Moderate Oil Price Rise
(quads)*

	1983	1990	2000	2010	2020
Africa	1.5	2.9	5.1	6.5	7.6
Asia	2.3	5.3	6.6	7.2	6.8
East Europe	21.2	30.7	37.5	39.4	41.9
JANZ	0.6	1.4	1.9	2.3	2.3
Latin America	2.6	3.6	6.0	6.4	5.2
Middle East	1.5	3.3	6.7	8.9	10.6
North America	19.5	25.4	27.4	29.5	33.1
West Europe	5.9	5.7	5.5	5.1	5.1
World Total	55.1	78.4	96.7	105.3	112.6

* Original figures were reported in exajoules. One exajoule equals approximately one quadrillion Btu.

Exhibit 5

International Gas Union

Gas Demand - By Region

Moderate Oil Price Scenario
(quads)*

	1983	1990	Potential Economic 2000	2010	2020
Africa	0.7	1.5	2.8	4.1	6.0
Asia	1.5	3.2	4.6	5.4	5.7
East Europe	19.5	24.7	34.2	37.0	40.0
JANZ	1.7	3.2	3.8	3.9	4.4
Latin America	2.4	3.0	4.2	5.2	6.3
Middle East	1.2	2.9	4.8	5.4	6.0
North America	20.5	24.1	26.5	28.6	31.2
West Europe	7.7	9.1	9.9	10.3	10.6
World total	55	72	91	100	110

more than twice the 1983 level. The projected potential gas demand is shown in Exhibit 5.

When comparing the potential production and consumption estimates to economically recoverable gas resources, the IGU report concluded:

"When the then remaining total reserves are related to the potential production in those years, these reserves would in the moderate case then still suffice for 100 and 70 years of production in 2000 and 2020 respectively, thus stretching the natural gas contribution to well into the next century."

U.S. Natural Gas Resources and Demand

U.S. Gas Supply

What about natural gas supply? To begin with the fundamentals, gas supply never should have been an issue. The real issue was always regulation. Regulation created the shortages and much of the acrimony on the interstate pipeline system in the 1970s. The supply in the ground is adequate for hundreds of years. When you compare the economically recoverable and usable coal and gas resource bases, they are roughly the same.

U.S. natural gas consumption in 1987 was 17 Tcf. The nation has discovered – and has available to be produced – about 160 Tcf of proven reserves in the lower 48 states. The Potential Gas Committee – the industry group that evaluates undiscovered gas resources – has estimated that there are an additional 620 Tcf of conventional gas resources that can be produced with foreseeable technology and economics. Therefore, conventional lower 48 states gas resources total nearly 800 Tcf – 160 Tcf of proved reserves and 620 Tcf of potential. This is equivalent to nearly 50 years of supplies at the current usage level.

The perception that the nation is using up this conventional recoverable resource base (the finite resource theory) is the rationale for projections of much higher gas prices beyond 2000. This theory is wrong. In addition to conventional gas resources, unconventional gas resources add 200 years of supplies. These unconventional sources include: eastern Devonian shales, western tight sands, coal-seam methane and enhanced gas recovery. Technology development will result in these vast unconventional supplies becoming conventional.

In addition to lower-48 states' gas, a number of gas supply sources will help satisfy domestic requirements as we move up the gas supply curve. Our assessment of lower-48 states' gas supply capability is supported by the replacement of production with new reserves, i.e., reserve additions, that have far exceeded the expectations of just a few years ago. In the period 1981-1986, aggregate reserve additions in the lower-48 states were 93 percent of production—with 1986 at 96 percent. These numbers prove that, if gas demand is there and the price is right, supplies will be discovered and available. What is interesting is that even when demand and prices are poor—as in 1986—gas supplies held up better than expected. The relatively strong showing for gas must be contrasted with oil which saw a sharp decline in domestic production and reserve replacement.

Until the last decade, the development of the natural gas resource base was a stepchild of oil exploration and development activities. This has changed somewhat—it will change dramatically over the next decade. The relative size of the undiscovered resource base, as well as the economics of exploration and production, will increasingly favor natural gas. By 1990, lower-48 states' natural gas production should be at least 50 percent greater than crude oil production. A survey of literature (e.g., annual reports) by major and independent oil companies confirms the view that the exploration and production industry is increasingly viewing gas as the fuel of the future.

Gas exploration success over the last five years further confirms our resource assessment. Recent gas discoveries have opened up new frontiers that will eventually add significantly to the resource base. For example, major new gas finds in the eastern and western Gulf of Mexico and deep offshore in the Gulf will likely stabilize production capability in this high productivity area into the foreseeable future. Exploration successes in frontier onshore areas have the promise of increasing future national production capability.

A.G.A. has forecast that excess domestic production capability ("gas bubble") will be gone by 1990. There has been all too much emphasis on the gas bubble—for the wrong reasons. Some people think the end of the bubble will mean shortages and much higher prices. Others have overlooked the negative impact of the bubble on gas supply activity, including reserve additions.

When the bubble ends the result will likely be a smooth return to balance—not shortages. The rig count has started up, and drilling will be stimulated as the bubble gradually diminishes. A.G.A. analysis indicates that, as the bubble contracts, several additional sources of gas can become rapidly available to ensure that no deliverability shortage occurs. These include such sources as presently uncommitted, nonproducing gas reserves (estimated by EIA as 9 Tcf, and not part of the bubble), gas which could be available from infill drilling, and Canadian gas. Prices may rise, but the risk of significant, sustained deliverability problems seems to us quite low. Other factors help ensure against a shortage as occurred in the winter 1976/77. For example, most gas-using industrial facilities and powerplants can now utilize at least one other fuel as a "backup." Today, 5-6 Tcf of demand can drop off the system if gas prices rise a few cents. And this is a choice made by the users. It has never been so easy to shift gas quickly through the nationwide million-mile transmission/distribution network – by sales, carriage, or both – from areas of surplus to areas of need. The flexibility of the gas delivery system has improved dramatically over the last decade.

As a result of these gas supply developments, estimates of the quantities of natural gas that potentially will be available to the marketplace under foreseeable conditions in this century have increased. In particular, the A.G.A. Gas Supply Committee now estimates that lower-48 gas wells, including gas from tight formations, will be producing between 14.8 and 23.5 Tcf annually in the year 2010 (see Exhibits 6-8). In other words, even without Canadian gas, domestic natural gas supplies will likely be adequate to maintain current consumption, and could support as much as a 40 percent increase. When all supply sources are considered, the committee now projects year 2010 supply capability at 18.9-28.2 Tcf, again compared to consumption of about 17 Tcf in 1987. [4]

Exhibit 6

U. S. Gas Supply - Low Price Scenario4/
(in Tcf)

	Actual	1990	2000	2010
Oil Price Assumption ($1987/bbl)	$18	$16	$22	$29
Gas Price Assumption (1987/MMBtu)	$1.85	$1.65	$2.95	$4.75
Lower-48				
Conventional	15.7*	16.0	13.5	11.5
Incremental Sources**	----	***	1.6	3.3
Imports				
Canadian	0.7	1.5	1.7	1.7
Mexican	----	----	1.0	1.2
LNG	***	***	0.2	0.4
Alaskan Gas Pipeline	----	----	----	0.7
Other				
Synthetics and Nonconventional	***	***	***	***
Totals(rounded)	16.5*	17.6	18.0	18.9

*Lower-48 Production Capability estimated to be 19.0 Tcf in 1986.
**Incremental Sources include gas from tight formations, coal seams, and co- production.
***Less than 100 Bcf.

Exhibit 7

U. S. Gas Supply - High Price Scenario4/
(in Tcf)

	Actual	1990	2000	2010
Oil Price Assumption ($1987/bbl)	$18	$20	$28	$40
Gas Price Assumption (1987/MMBtu)	$1.85	$2.05	$3.60	$6.20
Lower-48				
Conventional	15.7*	17.0	16.0	14.0
Incremental Sources**	----	0.4	2.7	4.6
Imports				
Canadian	0.7	1.5	2.0	2.5
Mexican	----	----	0.5	0.6
LNG	***	0.2	0.4	0.7
Alaskan Gas				
Pipeline	----	----	0.7	1.2
Other				
Synthetics and Nonconventional	***	***	0.1	0.2
Totals(rounded)	16.5*	19.2	22.4	23.8

*Lower-48 Production Capability estimated to be 19.0 Tcf in 1986.
**Incremental Sources include gas from tight formations, coal seams, and co-production.
***Less than 100 Bcf.

Exhibit 8

U. S. Gas Supply-OPEC Dominance Scenario4/
(in Tcf)

	Actual	1990	2000	2010
Oil Price Assumption ($1987/bbl)	$18	$20	$35	$60
Gas Price Assumption (1987/MMBtu)	$1.85	$2.05	$4.55	$9.30
Lower-48				
Conventional	15.7*	17.0	19.0	16.0
Incremental Sources**	----	0.4	3.7	7.5
Imports				
Canadian	0.7	1.5	2.0	2.5
Mexican	----	----	----	----
LNG	***	0.2	0.7	0.7
Alaskan Gas				
Pipeline	----	----	0.7	1.2
Other				
Synthetics and Nonconventional	***	***	0.2	0.3
Totals(rounded)	16.5*	19.2	26.2	28.2

*Lower-48 Production Capability estimated to be 19.0 Tcf in 1986.
**Incremental Sources include gas from tight formations, coal seams, and co-production.
***Less than 100 Bcf.

U.S. Gas Demand

Natural gas demand in the U.S. has proved to be highly price elastic. U.S. gas sales today are 25 percent lower than they were in the early 1970s. Fortunately, a greatly increased competitiveness and new end-use technologies are creating new prospects for natural gas use.

The A.G.A. Gas Demand Committee has estimated that these new markets could potentially add 1.7 - 7.4 Tcf of annual demand by 2010. [5,6]. (See Exhibit 9).

Exhibit 9

U. S. New Natural Gas Market Demand Forecasts5/
(Tcf)

	Business as Usual	Year 2000 Attainable	Potential
Cooling	0.09	0.37	0.79
Cogeneration	0.80	1.20	1.70
Combined Cycle	0.30	0.60	1.20
Select Use	0.10	0.30	1.00
CNG Vehicles	0.01	0.11	0.32
Total	1.30	2.58	5.01

Environmental and Efficiency Advantages of Natural Gas

Environmental impacts are of concern both at the point of end-use energy consumption and at all points in the chain ("full cycle") of production, processing and distribution. The environmental advantages of natural gas relative to other energy forms are equally impressive from either perspective.

Full-Cycle Emissions

For example, a recent study of pollution associated with residential sector energy consumption in the United States found that, when emissions from the full energy cycle are considered, use of natural gas space heating and other appliances results in only 15 to 20 percent of the total air emissions, and less than one percent of both the total water pollutants and noncombustible solid wastes compared with electric appliances.

Natural Gas Combustion

From the end-use or power plant combustion perspective, natural gas is also inherently cleaner than the combustion of other fossil fuels. [7].

Carbon Dioxide

Use of natural gas in lieu of coal or oil can significantly reduce carbon dioxide emissions. In boilers, natural gas emits roughly half as much carbon dioxide per unit of energy as coal and 30 percent less than oil. [8,9]. In combined-cycle units, gas emits only 36 percent as much carbon dioxide as a coal-fired boiler. In vehicular applications the benefit may be greater. [10].

Sulfur Dioxide

Sulfur dioxide emissions from natural gas combustion are virtually zero. A boiler operating on natural gas emits, on average, only 0.0006 pounds of SO_2 per million Btu (MMBtu) of fuel burned. In contrast, "typical" grades of fuel oil and coal consumed in large boilers (e.g., 1.5 percent sulfur oil; 2 percent sulfur eastern coal, uncontrolled) emit 1.6 pounds of SO_2 per MMBtu and 3.6 pounds per MMBtu, respectively. [11].

Particulate Matter

Natural gas combustion emits almost no particulate matter. The burning of natural gas results in particulate emissions averaging only 0.01 pounds per MMBtu of fuel consumed. In contrast, the combustion of 1.5 percent sulfur fuel oil emits 0.12 pounds of particulate matter per MMBtu, and eastern coal with a 10 percent ash content emits 7.60 pounds of particulates per MMBtu (uncontrolled). [12]. Various control devices are used to limit particulate emissions from large coal-burning facilities. For example, electrostatic precipitators can reduce the particulate emissions at a large (500 megawatt) powerplant by over 98 percent at a capital cost in the range of $50 million.

Carbon Monoxide

Nearly three-fourths of all carbon monoxide emissions stem from the operation of highway vehicles. The use of natural gas as a vehicular fuel can significantly reduce (and in many cases virtually eliminate) carbon monoxide emissions. [13].

Hydrocarbons

The world natural gas industry is both a source of atmospheric methane, as when gas is vented during oil production, and a means of reducing methane emissions, as when coal mine or landfill methane is captured and utilized. However, atmospheric methane attributable to the natural gas industry is negligible. According to a paper entitled "Trace Gas Scenarios" presented at the UNEP Economic Workshop on the Ozone Layers in May of 1986, all human activities and fossil fuel usage are a source of only about 5 percent of atmospheric methane loadings. The remaining 95 percent is attributable to rice paddies (38 percent), wetlands (34 percent), enteric fermentation (e.g., cattle and sheep, 20 percent), and fresh-water lakes, oceans and tundra (3 percent).

In the U.S., the sum of unaccounted for gas and vented and flared gas comprised 2.5 percent of total gas production in 1985. It is impossible to say with certainty what portion of this 2.5 percent represents methane released into the atmosphere, but it is likely less than half or 1.25 percent of total production. [14].

Gasoline- and diesel-powered transportation vehicles and various petroleum operations are responsible for the majority of man-made hydrocarbons emissions. The conversion of a gasoline-powered vehicle to natural gas can reduce reactive (non-methane) hydrocarbon emissions by as much as 92 percent. [13]. Total tailpipe hydrocarbon emissions (including methane) may, with today's technology, equal total tailpipe hydrocarbon emissions from a gasoline vehicle, but evaporative and filling emissions (the source of 65 percent of vehicle associated HC emissions) are essentially zero. As systems designed and tuned for efficient natural gas operation become more common, even these low methane emissions will be reduced.

Nitrogen Oxides

NO_x emissions from large boilers operating on coal may be over three times as great as emissions from gas-fired boilers. Depending on the coal and boiler type, NO_x emissions from large coal-fired boilers range from 0.9 to 2.6 pounds per MMBtu of fuel combusted. In contrast, comparably sized gas-fired units emit only about 0.7 pounds of NO_x per MMBtu. Large oil-fired boiler NO_x emissions are approximately equal to, or somewhat greater than, gas-fired boiler NO_x emissions. [15]. Furthermore, over 65 percent of all nitrogen oxide emissions stem from highway vehicles and large coal-fired facilities. Vehicles fueled with natural gas may emit as little as 35 percent as much NO_x as do gasoline-powered vehicles, depending upon vehicle design, tuning, etc. [13].

Chlorofluorocarbons

Chlorofluorocarbons (CFCs) have been identified as a major contributor to the breakdown of the Earth's protective ozone layer. Most gas cooling systems are based on technologies that do not use chlorofluorocarbons, while electric systems rely on CFCs as "coolants."

Non-Combustible Solid Waste and Sludge

Natural gas combustion results in no ash or sludge production. According to the U.S. Environmental Protection Agency, a 1,000 MW powerplant operating on high-sulfur coal with flue gas desulfurization would produce 700,000 tons of sludge per year and 250,000 tons of ash. A unit of the same size operating on 3 percent sulfur residual oil with a scrubber would produce no ash, but would produce 450,000 tons of sludge annually. (Oil desulfurization at the refinery is an option to scrubbing at the plant site.) There would be no production of either sludge or ash if the plant were to operate on natural gas. [16].

Dissolved Solids

Water-dissolved solids discharges from natural gas industry operations are negligible. Coal mining, processing and combustion account for approximately one-half of all dissolved-solids discharges in the U.S., versus less than 5 percent attributable to natural gas operations. [17].

Water Consumption

Water consumption associated with natural gas use is generally far lower than for other energy alternatives which require greater processing or conversion. Even when natural gas is used for electric power generation using combined-cycle technology, total treated water consumption by the gas-fired plant is only half that associated with coal-fired electricity generation. [18].

Capital Efficiency and Resource Advantages of Natural Gas

As a rule, supply and utilization systems based on gaseous fuels require substantially less initial capital investment than alternative energy systems. For example, supplying added quantities of gaseous fuel from conventional lower-48 gas resources for residential spaceheating would require only 29 percent to 46 percent of the initial capital required for the same amount of useful heat provided by nuclear or coal-fired electric power. [19].

A major factor contributing to the capital efficiency of gas systems is their high thermal efficiency. For example, a unit of natural gas can be produced and delivered to the end-user at an efficiency of nearly 89 percent. By comparison, when coal is used to generate electricity, the efficiency of generation and delivery is only 26-28 percent – that is, more than 70 percent of the resource energy is wasted. Even when a high-efficiency electric heat pump is used, the total efficiency of the electric cycle is 48 percent as compared to 85 percent for the gas system using a high-efficiency furnace.

New Natural Gas Applications and Technologies

For the first time in over 15 years, the U.S. gas industry – which accounts for one-third of world gas production – has adequate supplies available at prices which will permit effective competition in all major energy markets. The resulting stimulus to gas technology and market development in the U.S. is affecting gas technology worldwide.

This section of the paper provides an overview of some of the major gas utilization applications and technologies which are emerging, with particular emphasis upon environmental benefits.

Natural Gas-Fueled Combined-Cycle Electric Power Generation

Whereas two-thirds of the energy input to conventional fossil fuel boiler systems is normally lost to the environment (30 to 34 percent efficiency), the waste heat from the gas turbine in a combined-cycle unit is captured and utilized. Thus, combined-cycle system efficiency is in the 45 percent range or above.

A major advantage of combined-cycle systems is their relatively low capital cost – approximately one-third the cost per kilowatt of coal-based capacity, and about one-fifth the cost per kilowatt of nuclear capacity. The cost of combined-cycle systems, most of which are largely pre-packaged, has risen far less rapidly than coal and nuclear units. A recent study found that in the United States a new 240 megawatt (MW) combined-cycle unit produces electricity at a cost which is only 54 to 92 percent of the cost of electricity produced by a new, comparably sized coal unit.

In addition, combined-cycle systems can be brought on-line quickly and in stages. For example, two-thirds of the electric capacity of a 100 MW combined-cycle system (the gas turbine component) can be on-line in 12 to 18 months, with the balance operating in an additional 6 months. In contrast, new coal or nuclear units require approximately 8-15 years for design and construction.

The environmental advantages of combined-cycle plants are also significant. For example: (1) a gas-fired combined-cycle unit emits less than 1 percent of the SO_2 and particulates associated with an equivalent coal-fired unit; (2) emissions of NO_x from the combined-cycle plant are 57 percent of the coal unit emissions; (3) the coal plant would produce 60 thousand tons of ash per year and 168 thousand tons of sludge while combined-cycle units produce neither of these non-combustible solid wastes; and (4) treated water consumption by the combined-cycle plant would be 15 million gallons per year, one-half the consumption of the coal plant. [20].

For example, if 100 electric power plants totaling 24,000 MW were to be built instead of the same amount of new coal-fired electric capacity with full NSPS controls, annual SO_2 emissions would be 410,000 tons lower and annual NO_x emissions would be reduced by 105,000 tons (-43 percent).

- If the same 100 new gas-fired powerplants were to displace existing uncontrolled coal-fired powerplants, annual SO_2 emissions would be reduced by 3.9 million tons, while NO_x emissions would be reduced by up to 545,000 tons annually (-80 percent).

- Total annual natural gas use by these 100 plants would be about 1 Tcf.

- Annual CO_2 emissions would be 143,000 tons if the plants were coal-fired, compared to 51,000 tons for combined-cycle gas (see Exhibit 10).

Natural Gas-Fired Cogeneration

Whereas combined-cycle units (see above) utilize waste heat to generate electricity, cogeneration systems use waste heat for a variety of purposes – e.g., space or water heating and industrial process needs.

Gas-fired cogeneration systems are an attractive option from both an environmental and an energy efficiency standpoint. According to the U.S. Office of Technology Assessment, a gas turbine cogeneration system would require approximately 25 percent less input energy than the combination of a new coal-fired electric powerplant and an

Exhibit 10

Emissions Resulting From 100 New 240 MW
Gas Combined Cycle Powerplants Versus Coal
(Annual Tons)

	(A) Combined Cycle	(B) NSPS Coal	Difference (B-A)	(C) Uncontrolled Coal	Difference (C-A)
Sulfur Dioxide	0.3	410	410	3,900	3,900
Particulate Matter	0.9	21	20	1,700	1,700
Nitrogen Oxides	135	240	105	680	545
Carbon Dioxide	51,000	143,000	92,000	143,000	92,000
Total Gas Input/Year		1.08 Quad			

oil-fired boiler producing steam (assuming optimal system sizing). In addition, the gas-fired cogenerator would emit less than 1 percent of the SO_X, 27 percent of the particulates, and 50 percent of the NO_X produced by the comparably sized conventional coal- and oil-based system (with pollution control equipment).

Gas-fueled cogeneration is also economically and environmentally preferable to coal-fueled cogeneration in large-scale industrial and commercial applications. For example, when compared to a coal-fueled system with a scrubber and an electrostatic precipitator, a gas-fueled steam turbine emits only 6 percent of the sulfur oxides, 10 percent of the particulates, and 28 percent of the nitrogen oxides. [21,22].

Because of these advantages, 45 percent of the existing cogeneration capacity and 66 percent of the capacity on order or planned in the U.S. are gas-fired. [23].

Natural Gas Repowering

A significant portion of U.S. electric utility generating capacity is approaching the end of its design life, which is roughly 30 to 35 years. By 1990, approximately 25 percent of all fossil-fueled generating capacity will be at least 30 years old. As generating equipment ages, it presents plant operators with a variety of problems, including reduced efficiency and reliability. A 30-year-old unit requires 6 to 8 percent more fuel—at a cost of millions of dollars annually—to generate the same amount of power as it did when it was first put into service. Unit reliability also suffers over time. A 10-year-old coal-fired plant has a forced outage rate of about 5 percent; however, this rate increases to nearly 15 percent within 30 years.

Combined-cycle repowering refers to the integration of new and used equipment at an existing site, with the final equipment configuration resembling a new gas-fired combined-cycle unit.

The type of repowering will vary from site to site. For example:

1) Peaking turbine repowering refers to the addition of a steam turbine and heat-recovery unit to an existing gas turbine, with the efficiency improvement allowing the unit to convert from peaking to baseload operation.
2) Heat-recovery repowering is the replacement of an old coal boiler with a new gas turbine and heat-recovery unit, leaving the existing steam turbine in place.
3) Boiler repowering feeds the exhaust from a new gas turbine into an existing coal boiler, replacing the existing coal boiler's forced draft fans and combustion air heaters, and significantly reducing the system's electrical and fuel consumption.

The cost of producing electricity in a repowered unit is only about 60 percent of the cost of producing electricity in a new, coal-fired power plant. Even if the delivered price of natural gas is substantially higher than the price of coal, this price differential is likely to be more than offset by the lower capital and operating costs and maintenance expenses associated with the repowered system. The capital charge for a repowered unit is very low—between 0.6 cents and 1.0 cents per kilowatt hour (kWh), versus 3.3 cents/kWh for a new coal unit—since a repowered unit uses some existing equipment and does not require pollution-control devices. A related benefit is the ability to keep the power-generating facility on the existing site, eliminating the need for acquiring additional land, rights-of-way and various permits required for new construction. [23].

Another important economic factor is that the fuel cost advantage enjoyed by coal units is further diminished because of the 10 to 30 percent higher efficiency of repowered gas units.

Similarly, the operating and maintenance expenses of repowered units are only about one-third that of new coal plants, which require coal storage and handling, as well as the maintenance of pollution-control equipment and sludge and ash disposal.

Fuel Switching and Select Use in Boilers

Switching to gas can be a highly attractive and cost-effective control technique in existing boilers.

- Using gas in conjunction with, or in place of, coal or residual oil entails minimal up-front capital costs and relatively low life-cycle costs. Many large coal and oil boilers currently have the capability to also burn gas. Those that do not can add this capability at a very low cost—$5 to $20 per kW—with little or no loss in boiler rating or efficiency. [25,26].

- The cost of removing SO_2 via "select use of gas" ranges from $50 to over $1,000 per ton, often in the $200 to $400 range. Alternative control strategies range from $250 to $2,000 per ton removal. NO_x control may be thought of as a free extra gas benefit.

- As noted above, switching to gas reduces a number of pollutants simultaneously. In addition, it offers quick response and supply security.

- Gas can also be used in conjunction with other control strategies, such as high-sulfur coal washing, thereby allowing the continued use of local coals. For example, SO_2 reductions of 50 to 66 percent could be achieved while maintaining the local coal as 70 percent of the total fuel mix.

A detailed study by A.G.A. and ANR Pipeline Company of 43 major SO_2 emitters in Wisconsin confirms the potential of select use. [27].

- Select use was considered to be very attractive for 16 of the 43 plants. These plants could burn gas for SO_2 control with no expenditures for pipeline additions or boiler modifications.

- Annual SO_2 emissions could be reduced by 134,000 tons per year (roughly 50 percent for the 16 plants) at a cost of $156 to $440 per ton removed. This amount is equivalent to one-fourth the total emissions of all 43 plants.

The substitution of 1 Tcf/year of natural gas for high sulfur coal in boilers would have a significant impact on a number of pollutants.

- SO_2 and NO_x emissions would be reduced by 2.8 million tons and 0.5 million tons per year, respectively. By combining select use with coal washing, the SO_2 reduction could be roughly doubled.

- In addition, non-reactive hydrocarbons would be reduced by 2.3 thousand tons per year, and CO would be reduced by 5 thousand tons per year.

A recent A.G.A. study also investigated the relative advantages of natural gas boilers and fluidized bed coal boilers. [28].

For a typical industrial application, the use of natural gas in a boiler costs only 73 percent of the cost of burning coal in an atmospheric fluidized bed combustion (FBC) boiler, while emitting only a fraction of the air and solid waste pollution used by the FBC system. While the fuel costs of the gas system are higher, the gas boiler's lower capital and non-fuel operation and maintenance expenses more than offset the fuel cost advantage of the FBC system.

Eight alternate cases evaluated economic sensitivity of changing various parameters, including: unit size, fuel prices, capacity utilization, discount rates, and capital costs. In all cases, the gas option was found to cost less, ranging from 64 percent to 86 percent of the FBC unit. However, the option of burning coal in a conventional boiler equipped with a scrubber was found to be more costly than the FBC boiler.

Although the combustion process of an FBC boiler reduces the sulfur dioxide and nitrogen oxides emissions associated with uncontrolled coal burning, and usually employs a fabric filter to control particulate emissions, air pollution emissions are much greater for the FBC boiler than the gas boiler. For example, the natural gas system emits less than one percent of the sulfur dioxide, only 24 percent of the nitrogen oxides and 11 percent of the particulate matter compared to a similarly sized coal FBC.

Cofiring

Gas cofiring involves simultaneously burning natural gas—typically comprising less than 30 percent of boiler fuel input—with coal or oil in the same boiler to reduce emissions and improve operational performance.

Cofiring gas with coal in the same boiler offers a variety of environmental and operational benefits. Not only does the use of gas reduce emissions of SO_2, NO_x, particulate matter, sludge, ash and many other pollutants, it also can enhance boiler performance, resulting in lower maintenance costs and fewer plant outages. Cofiring also provides cleaner and quicker start-ups.

Recent field experience has documented a greater than one-to-one reduction in SO_2 and NO_x emissions with cofiring. Theoretically, for SO_2, the reduction should be directly proportional to the amount of coal displaced. Using gas to provide 10 percent of the fuel input, for example, should result in 10 percent less SO_2. However, in practice, GRI research showed that measured levels reflected significantly greater reduction, apparently due to the tendency of the sulfur to bind to the ash during the process. Related tests have shown that NO_x emissions are reduced at a rate roughly three times greater than the proportion of gas used. [29].

Seasonal Gas Substitution

Natural gas also can be used to reduce SO_2 emissions by substituting or alternating its use with coal or high-sulfur oil. Controlling SO_2 emissions during specific periods of the year may be more cost-effective than year-round controls, as indicated in studies conducted by the Massachusetts Institute of Technology (MIT). According to MIT, 70 percent of annual acid deposition nationwide falls from May through October, with 54 percent falling during the four months of June through September. This coincides with the period of greatest available natural gas pipeline capacity. Substituting gas for dirtier fuels also can provide very cost-effective compliance under EPA's emissions trading program. [26,30].

Gas Reburning with Sorbent Injection

In gas reburning, natural gas replaces a portion of the coal burned in a boiler. The gas, injected into the upper region of the furnace, creates a fuel-rich zone in which NO_x is chemically changed into a harmless molecular nitrogen. Research conducted by GRI and Energy and Environmental Research Corp. shows that reburning could reduce NO_x emissions by 50 to 60 percent. [31].

Sorbent injection involves mixing a sulfur-absorbing chemical with the coal to remove the sulfur oxide gases that are formed when coal burns. According to Energy and Environmental Research, use of sorbent injection is expected to cut SO_2 emissions by at least one-half.

Vehicular Applications

Conversion of gasoline or diesel fueled vehicles to natural gas can reduce carbon monoxide emissions by up to 99 percent, reduce nitrogen oxides emissions by up to 65 percent, and/or reduce reactive hydrocarbon emissions by up to 92 percent. This is particularly true for trucks, some buses, and older cars, which account for 60-90 percent of vehicular emissions. Actual emissions reductions achieved depend upon a large number of factors including tuning, age and design of the vehicle, and the condition and technology of the gasoline emissions controls and natural gas conversion kit. [13].

- Conversion of 1 million cars, less than 10 percent of U.S. fleet vehicles, to natural gas could reduce carbon monoxide emissions by up to 223,000 tons/year, reduce nitrogen oxides emissions by up to 27,000 tons/year and/or reduce hydrocarbon emissions by up to 44,000 tons/year. These vehicles would use less than 100 Bcf of natural gas per year – one-half of one percent of annual U.S. natural gas production.

- Stated alternatively, for each 1 trillion cubic feet (Tcf) of natural gas used annually in vehicles, carbon monoxide emissions could be reduced by up to 2.2 million tons annually (a four percent national reduction), nitrogen oxides emissions could be reduced by up to 270,000 tons annually (a one percent national reduction), and hydrocarbon emissions could be reduced by up to 440,000 tons annually (a two percent national reduction). Reductions of this magnitude, if focused in ozone nonattainment areas, could allow some areas to come into compliance.

The cost of these emission reductions may vary from free (where fuel and maintenance cost savings offset capital costs of conversion) to several thousand dollars per ton. Reduction of other emissions may be thought of as a bonus.

Residential Gas Systems

Natural gas has long been recognized to be the most economical fuel for residential space heating, water heating, clothes drying and cooking. Recent commercialization of a new generation of highly efficient gas appliances has enhanced this advantage. [32,33].

It is less often recognized that, as with other gas uses, residential gas systems are also best from the perspective that they utilize less primary energy and result in less environmental pollution than do most alternatives.

Natural Gas Cooling Systems ("Air Conditioning")

The recent commercialization of new, high-efficiency gas cooling systems has greatly improved the attractiveness of natural gas for this application. American Gas Association analyses have demonstrated the cost advantages of gas cooling in many commercial applications. Perhaps even more important from an energy planning and environmental perspective, the availability of gas cooling may avoid the need for adding highly expensive electric peaking capacity to meet summer loads, while also balancing seasonal gas demand. [34,35].

Conclusion

Natural gas resources are abundant, widely distributed and highly underutilized. By all measures—air emissions, solid wastes, water pollution, and water consumption—natural gas is by far the most environmentally desirable fossil fuel. Compared to other energy alternatives natural gas systems are also highly capital and resource efficient, requiring significantly lower capital investment to produce, deliver and utilize a unit of energy. The emergence of highly attractive natural gas-fired electricity generation technologies has enhanced these advantages. Natural gas can and should play a major role in environmental protection in many nations.

References

1. Portions of this section are quoted with permission from Nelson E. Hay, *Energy Policy Strategies for Developing Nations: The Role of Natural Gas*, to be published.

2. American Gas Association, *A Comparison of World Remaining World Gas and Oil Resources* (Arlington, VA, December 17, 1982).

3. International Gas Union, *Report of Task Force: World Gas Supply and Demand, 1983-2000*, (Presented at the 16th World Gas Conference, Munich, Germany, March 1985).

4. American Gas Association, *The Gas Energy Supply Outlook: 1987-2010* (Arlington, VA, October 1987), pp. 4-6. 5. American Gas Association, The Outlook for Gas Demand in New Markets: 1986-2010 (Arlington, VA, August 1986), p. 11.

6. American Gas Association, *A.G.A.-TERA Analysis 88-1 Base Case* (Arlington, VA, January 15, 1988).

7. American Gas Association, *Fact Book: Energy, the Environment and Natural Gas* (Arlington, VA, October 1983).

8. American Gas Association, *Carbon Dioxide Emissions from Fossil Fuel Combustion and From Coal Gasification* (Arlington, VA, September 2, 1977).

9. Krickerberger, Kit R. and Lubore, Stephen H., *Feasibility of Utilizing Carbon Dioxide Produced During the High-Btu Coal Gasification Process for Enhanced Oil Recovery* (McLean, VA, MITRE Corp., August 1980).

10. See, for example, Gyushee, David E., *Carbon Dioxide Emissions for Methanol As a Vehicle Fue*l (Washington, DC, Congressional Research Service, January 4, 1988).

11. U.S. Environmental Protection Agency, Office of Air and Waste Management, *Compilation of Air Pollutant Emissions Factors* (Research Triangle Park, NC, August 1977), pp. 1.1-3, 1.3-2, 1.4-2.

12. U.S. Environmental Protection Agency, Office of Air and Waste Management, Op. Cit., pp. 1.13, 1.3-2, 1.4-2.

13. American Gas Association, *Ozone and Carbon Monoxide: The Role of Natural Gas in Attaining Clean Air Act Compliance: 1988 Update* (Arlington, VA, March 25, 1988).

14. American Gas Association, *Lost and Unaccounted For Gas* (Arlington, VA, December 5, 1986), p. 3.

15. U.S. Environmental Protection Agency, Office of Waste Management, Op. Cit., pp. 1.1-3, 1.3-2, 1.4-2.

16. U.S. Environmental Protection Agency, Research and Development—Energy Minerals and Industry, *Energy/Environmental Fact Book* (Washington, DC, March 1978) p. 24.

17. U.S. Department of Energy, Office of Environmental Assessment, *National Environmental Impact Projection II* (Washington, DC, March 1983), pp. 4-7.

18. General Electric, GER 3401, *STAG Combined Cycle Product Line and Performance Characteristics* (Schnectady, NY, 1984), p. 568316A.

19. American Gas Association, *Comparison of Capital Costs to Consumers of Alternative New Domestic Energy Supplies* (Arlington, VA, December 14, 1983).

20. American Gas Association, *Natural Gas-fueled Combined Cycle Electric Power Generation—An Attractive Option* (Arlington, VA, May 10, 1985).

21. American Gas Association, *An Economic and Environmental Comparison of Natural Gas and Coal Use for Large-Scale Industrial Cogeneration* (Arlington, VA, October 26, 1984).

22. Hay, Nelson E., Editor, *Guide to Natural Gas Cogeneration* (Lilburn, GA, The Fairmont Press, Inc., 1988).

23. American Gas Association, *Estimated Current Natural Gas Cogeneration Capacity: 1988 Update* (Arlington, VA, March 18, 1988).

24. American Gas Association, *Repowering With Natural Gas—An Electricity Generating Option* (Arlington, VA, October 17, 1986).

25. Fay, James A.; Golomb, Dan S. and Zachariades, Savvakis C., *Feasibility and Cost of Converting Oil- and Coal-Fired Utility Boilers to Intermittent Use of Natural Gas* (Cambridge, MA, Massachusetts Institute of Technology, December, 1986).

26. Fay, James A. and Golomb, Dan S., *Economics of Seasonal Gas Substitution in Coal- and Oil-Fired Power Plants* (Cambridge, MA, Massachusetts Institute of Technology, December 1987).

27. Armiak, Michael J., Correspondence to McKinney, Chris, Wisconsin Department of Natural Resources (Detroit, MI, ANR Pipeline Co., October 29, 1985).

28. American Gas Association, *Industrial Power Production Options -- A Comparison of Natural Gas Boilers and Coal-Fired Fluidized Bed Boiler*, (Arlington, VA, October 18, 1985).

29. Booth, R.C.; Breen, B.P.; Gallaer, C.A. and Glickert; R.W., *Energy Systems Associates, Natural Gas/Pulverized Coal Cofiring Performance Testing at an Electric Utility Boiler* (Chicago, IL, Gas Research Institute, July 1987).

30. Wilkinson, Paul L., *Seasonal Fuel Switching to Natural Gas—A Logical Approach to Reduce Sulfur Dioxide and Other Air Emissions* (A Presentation to the Air Pollution Control Association, Detroit, MI, June 16, 1985).

31. Ban, Stephen D.; Berkau Eugene E. and Pratapas, John M., *New Gas Combustion Technologies for Emission Control*, (A Presentation to the Conference on Select Use, Washington, DC, August 1985).

32. American Gas Association, *Consumer Costs of Natural Gas, Fuel Oil and Electricity in New Homes* (Arlington, VA, March 7, 1986).

33. American Gas Association, *Cost Advantages of Natural Gas in Existing Homes* (Arlington, VA, April 4, 1986).

34. American Gas Association, *A Life-Cycle Cost Analysis of the Market Potential of Residential Gas Air Conditioning* (Arlington, VA, November 30, 1984).

35. American Gas Association, *Analysis of Commercial Gas Cooling: 1986 Update*.

CONFERENCE PARTICIPANTS

Ms.	Amy	Abel
Congressional Research Service		
Mr.	J. Patrick	Adcock
Office of Technology Assessment		
Mr.	Sam	Altshuler
PG & E		
Mr.`	Richard	Artz
NOAA Air Resources Lab		
Mr.	Robert B.	Asplund
Virginia Power		
Ms.	Susan	Babb
Pace University School of Law		
Mr.	John	Bachmann
EPA		
Mr.	Thomas	Barlow
Florida Power & Light Company		
Mr.	Keith	Belton
American Chemical Society		
Dr.	Nicholas	Berenyi
Westvaco Corp.		
Mr.	Martin	Bernard
Argonne National Laboratory		
Mr.	Roger	Blair
EPA		
Mr.	Cary	Bloyd
Argonne National Laboratory		
Mr.	James	Bonevac
Va. State Corp. Commission		
Mr.	Michael	Bonoff
Int'l Science & Tech		
Mr.	Alvan	Bruch
Tennessee Valley Authority		
Mr.	Richard	Burkhart
U.S. EPA		
Mr.	Earl	Byron
Uni. of California, Div. of Environmental Studies		
Ms.	Ann	Carey
USDA Forest Service		
Mr.	Bill	Carr
Environment Canada		
Mr.	William J.	Carr
Ministry of the Environment		
Mr.	Douglas	Carter
Dept. of Energy		
Mr.	Edgar M.	Chase
Fairfax Co. Air Pollution Control		
Mr.	Jon	Clark
Environmental & Energy Study Institute		
Mr.	Gary	Cline
Pa. Electric Co.		
Ms.	Nancy	Cobb
NOAA		
Mr.	Peter E.	Coffey
New York Power Pool		
Mr.	Douglas	Cogan
Investor Responsibility Research Center		
Ms.	Anne E.	Colaccino
Yale University		
Mr.	Herman F.	Cole
Adirondack Park Agency		
Mr.	Richard	Cook
Va. Dept. of Air Pollution Control		
Ms.	Liz	Cook
Friends of the Earth		

	Martine	Costello
Transcript Telegram		
Ms.	Geraldine V.	Cox, Ph.D.
Chemical Manufacturers Association		
Ms.	Angela	Crump
US General A/C Office		
Mr.	Pat	DeLuca
American Industries & Resources Corp.		
Mr.	Jerome E.	Dobson
Marietta Energy Systems, Inc.		
Mr.	Dennis C.	Drehmel
EPA		
Mr.	Gene	Durman
U S EPA		
Mr.	Eric S.	Edgerton
Environmental Science & Engineering, Inc.		
Mr.	W. Ray	Effer
Ontario Hydro		
Dr.	Ivan	Fernandez
University of Maine		
Mr.	David	Filbert
Baltimore Co. Air quality mgmt		
Mr.	Richard	Flagler
Texas Agricultural Experiment Station		
Mr.	James	Flanagan
C-E Environmental		
Dr.	Carl A.	Fox
Desert Research Institute - Director of Bio. Sciences		
Ms.	Cari	Furiness
NCSU		
Mr.	Dan	Gealy
ARCO		
Mr.	Walter	Giles
Environment Canada		
Mr.	Ramesh V.	Godbole
Potomac Electric Power Company		
Mr.	Stewart	Goltz
University of Maine, Dept. Plant & Soil Sciences		
Mr.	Peter	Guerrero
US General A/C Office		
Ms.	Juliana	Haddad
Environment Canada		
Ms.	Elaine	Haemisegger
Temple,Barker & Sloane		
	Marcel L.	Halberstadt
Motor Vehicle Manu. Assoc. of the U.S., Inc.		
Mr.	Clifford A.	Haupt
U.S. Geological Survey		
Mr.	James	Healey
W.S.Flemling & Assoc.,Mt. Cloud Chemistry Project		
Mr.	Robert	Helie
Environment Canada		
Mr.	Donald	Henderson
National Park Service AIR		
Mr.	Eric	Hennen
Dairyland Power Cooperative		
	S.G.	Hildebrand
Martin Marietta Energy Systems, Inc.		
Ms.	Bette	Hileman
Press		
Mr.	Craig	Hillock
Life Systems, Inc.		
Mr.	W. Ted	Hinds
U.S. Dept. of Energy, Pacific NW Laboratory		

Mr.	Kenneth	Klenk
SAR		
Dr.	Richard L.	Klimisch
General Motors Corp.		
Mr.	Edward	Koubek
Chemistry Dept.		
Mr.	Robert T.	Lackey
U.S. Environmental Protection Agency		
Mr. Dennis	Dennis R.	Landsbug,Ph.D.
W.S. Fleming & Associates, Inc.		
Ms.	Jan	Lane
Dept. of Energy		
Mr.	John	Langstaaf
Alliance Technologies		
Dr.	George	Lauer
ARCO		
Dr.	Miriam	Lev-on
GE Environmental		
Ms.	Marie	Lihn
Gas Research Institute		
Mr.	Timothy	Lough
Va. State Corp. Commission		
Mr.	Dale	Lowry
National Climate Program Office		
Mr.	James	Lynch
Pa. State University		
Mr.	Jim	MaClean
Ontario Ministry of Natural Resources		
Ms.	Linda	MacQueen
Environment Canada		
Ms.	Kathleen	Manor
National Park Service		

Mr.	Robert	Honea
Marietta Energy Systems, Inc.		
Mr.	Peter	Horn
Mr.	Toru	Iwama
Georgetown Uni. Law Center		
Mr.	Art	Jaques
Environment Canada		
Mr.	Erik	Johnson
Greenpeace		
Dr.	Cameron C.	Johnson
Dept. of Commerce		
Mr.	Julian W.	Jones
US EPA		
Dr.	George	Jordy
Director for Program Analysis		
Ms.	Jodee	Joyce
National Science Foundation		
Ms.	Kimberly	Joyner
NCSU-Atmospheric Impacts Research Program		
Mr.	John R.	Justus
Congressional Reserach Service		
Mr.	Wayne	Kaiser
US Environmental Protection Agency, Region VII		
Mr.	Robert	Kane
Dept. of Energy		
Dr.	Bruce A.	Kenamore
Amoco Corporation		
Mr.	Richard	Kerr
Science		
Ms.	Katharine	Kimball
EPW Committee		

Mr. C. Robert Manor
Associate

Msd. Alice Mayer
Edison Electric Institute

Mr. Paul McArdle
American Gas Association

Mr. Robert McFadden
Motor Vehicle Manu. Assoc. of the U.S., Inc.

Dr. Mack McFarland
E.I. DuPont de Nemours & Co.

Mr. Jim McGranery

Mr. John M. McManus
American Electric Power Svc Corp.

Mr. Paul A. Michael,Ph.D
Brookhaven National Laboratories

Mr. Robert Missen
Pacific Power & Light

Ms. Lysa Modica
Alliance Technologies Corp.

Dr. Charles L. Mulchi
University of Maryland

Mr. Robert Myers
American Society of Agronomy

Dr. Norman Nielsen
Chrylser Motors Corporation

Pro Ray T. Oglesby
Dept. of Natural Resources

Mr. Bradford Owen
Consolidation Coal Company

Ms. Eustice Parnelle
Florida Power Corporation

Mr. Steve L. Pernick, Jr.
Duquesne Light Company

Mr. Tony Perpich
Minnesota Dept. of Public Service

Ms. Elizabeth Pezzullo
International Barometer

Mr. Steve Piccot
Alliance Technologies

Dr. Ranard J. Pickering
U.S. Geological Survey

Dr. Louis F. Pitelka
Electric Power Research Institute

Ms. Patricia Pitton
U.S. Environmental Protection Agency

Ms. Angela R. Powers
Dayton Power & Light Company

Ms. Theresa Pugh
National Association of Manufacturers

Mr. J.W. Ranney
US Dept. of Energy,(Marietta Energy Systems)

Ms. Lorraine Ratnik
International Relations, U.S. Group

Ms. Edith G. Read
Protection of the Adirondacks

Dr. Howard Reiquam
Gas Research Institute

Ms. Barbara Richman
Environment Magazine

Ms. Helen G. Rigg
Idaho Air Quality Bureau

Ms. Molly Ross
National Park Svc, Air Quality Div.

Ms. Diana Roth
Battele Pacific NW Laboratory

Ms. Sandy Rupp
US Export Council for Renewable Energy

Ms. Roslyn Scheid
Cornell University (undergraduate)

Dr. Fernando Schiappa
American Industries & Resources Corp.

Mr. Andy Schwarz
Temple, Barker & Sloane

Mr. Paul Schwengels
U S EPA

Leslie Sherman
NOAA

Mr Yasuhiro Shimizu
Environmental Attache

Ms. Margaret Singh
Argonne National Laboratory

Mr. John M. Skelly
Penn State University, Dept. of Plant Pathology

Mr. Albert Slechter
Chrylser Motors Corporation

Dr. Charles T. Smart
NJ Dept of Environmental Protection

Dr. Nancy Paige Smith
St. Mary's College of Maryland

J. R. Soulen
Tech and Mgmt Services

Mr. David South
Argonne National Laboratory

Mr. Don Spellman
Public Service Indiana

Ms. Teresa Spisak
US General A/C Office

Mr. Robert L. Srubar
E.I. Dupont de Nemours & Co.

Mr. Steve Stodola
Revet Laboratories

Mr. Larry Stowe
NCASC Training Staff

Dr. Joan Sturtevant
The University of Akron

Mr. Glenn Sutter
Royal Society of Canada

Ms. Rebecca Taggart
EPA REGION 3

Mr. George E. Taylor, Jr.
Oak Ridge National Laboratory

Mr. Donald F. Theiler
Wisconsin Dept. of Natural Resources

Ms. Clare Thorne
Senator Cranston's office

Ms. Susan Troll
Energy Office

Ms. Terri Turner
NYSFG

James Vail,Ph.D.
American Petroleum Institute

Mr. James G. Watkins
National Park Svc.,Dept. of the Interior

Mr. Charles Webb
Nat'l Council of the Paper Industry

Mr. Bill Wehrle
ERC-International

Mr.	Daniel	Weiss
Dairyland Power Cooperative		
Mr.	Melvin J.	Wentland
St. John Fisher - Biology Department		
Mr.	Marchant	Wentworth
Izaak Walton League of America		
Ms.	Carol	Werner
Environmental & Energy Study Institute		
Mr.	Karl	Wilkins
State of Maine		
Mr.	W.R.	Woodall, Jr.
Georgia Power Company		
Dr.	James N.	Woodman
North Carolina State University		
Mr.	Martin	Yerg
NCASC Training Staff		

CONFERENCE PROGRAM

Tuesday, November 15, 1988

Registration

Welcome
Elizabeth Thorndike, Executive Director, Center for Environmental Information, Inc.

THE GLOBAL CLIMATE

Moderator: **Edith Brown Weiss**, Professor of Law, Georgetown University Law Center

Global Climate Linkages
J. Christopher Bernabo, President, Science & Policy Associates, Inc.

The Climate System
Wallace S. Broecker, Newberry Professor of Geology, Columbia University

Break

Effects of Climate Change
Peter Robinson, Associate Professor, Department of Geography, University of North Carolina
Stephen P. Leatherman, Director, Laboratory for Coastal Research, University of Maryland

Luncheon

LINKAGES BETWEEN CLIMATE CHANGE AND OTHER ATMOSPHERIC ISSUES

Moderator: **James R. Mahoney**, Director, National Acid Precipitation Assessment Program

Acid Rain
Hans Martin, Senior Advisor, Environment Canada

Stratospheric Ozone
Robert C. Worrest, Manager, Stratospheric Ozone Research Program, U.S. Environmental Protection Agency

Break

Ambient Air Quality
Perry J. Samson, Associate Professor of Atmospheric Sciences, University of Michigan

Panel: Discussion, Questions and Answers
Hans Martin, Robert Worrest, Perry Samson

Adjourn

Wednesday, November 16, 1988

OPTIONS AND STRATEGIES

Moderator: **Rosina Bierbaum**, Project Director, Climate Change Assessment, U.S. Congress Office of Technology Assessment

Global Energy Use: A Quantitative Analysis
Richard Wilson, Mallinckrodt Professor of Physics, Harvard University

Control of Emissions
Ian Torrens, Director of Environmental Control Systems, Coal Combustion Systems Division, Electric Power Research Institute

Break

Energy Conservation and Alternate Energy Use
William R. Moomaw, Director, Program in Climate Change, Energy and Pollution, World Resources Institute

Environmental Modification
Greg Marland, Environmental Science Division, Oak Ridge National Laboratory

Luncheon

RESEARCH AND POLICY LINKAGES

Moderator: **J. Christopher Bernabo**, President, Science & Policy Associates, Inc.

Research Panel
Peter Jutro, Director, Marine Freshwater and Modeling Team, U.S. Environmental Protection Agency
Lester Machta, Director, Air Resources Laboratory, National Oceanic and Atmospheric Administration
John S. Perry, Staff Director, Committee on Global Change, National Academy of Sciences
Ralph M. Perhac, Director, Environmental Science Department, Electric Power Research Institute

Break

Policy Panel
John Hoffman, U.S. Environmental Protection Agency
Edward Williams, U.S. Department of Energy
Lester Lave, James Higgins Professor of Economics, Carnegie-Mellon University
Joseph Goffman, Staff Attorney, Environmental Defense Fund
Michael I. German, Vice President, Policy Evaluation and Analysis, American Gas Association

Adjourn